Calcium and Magnesium in Groundwater

Selected papers on hydrogeology

21

Series Editor: Dr. Nick S. Robins
Editor-in-Chief IAH Book Series, British Geological Survey, Wallingford, UK

INTERNATIONAL ASSOCIATION OF HYDROGEOLOGISTS

Calcium and Magnesium in Groundwater

Occurrence and significance for human health

Editor

Lidia Razowska-Jaworek
Polish Geological Institute-NRI, Sosnowiec, Poland

CRC Press
Taylor & Francis Group
Boca Raton London New York

CRC Press is an imprint of the
Taylor & Francis Group, an **informa** business

A BALKEMA BOOK

CRC Press
Taylor & Francis Group
6000 Broken Sound Parkway NW, Suite 300
Boca Raton, FL 33487-2742

First issued in paperback 2019

ISBN-13: 978-1-138-00032-2 (hbk)
ISBN-13: 978-0-367-37862-2 (pbk)

Library of Congress Cataloging-in-Publication Data

Calcium and magnesium in groundwater: occurrence and significance for human health / editor: Lidia Razowska-Jaworek. -- Edition 1.
    pages cm. -- (Selected papers on hydrogeology; 21)
Includes bibliographical references and index.
ISBN 978-1-138-00032-2 (hardback : alk. paper) 1. Groundwater--Pollution.
2. Calcium--Physiological effect. 3. Magnesium--Physiological effect.
I. Razowska-Jaworek, Lidia. II. Series: Hydrogeology (International Association of Hydrogeologists); v. 21.

TD426.C33 2014
551.49--dc23

                                    2014010028

Visit the Taylor & Francis Web site at
http://www.taylorandfrancis.com

and the CRC Press Web site at
http://www.crcpress.com

Table of contents

Preface

This volume is a set of chapters that focus on the occurrence and distribution of calcium and magnesium in groundwater, a topic not previously addressed in the thematic literature. However, this is not the only important feature of this book. It also has an important multidisciplinary character as its main objective is to present results of studies on calcium and magnesium in groundwater, along with the significance of these constituents for human health. Why just calcium and magnesium? The answer is simple – because these are the most important ionic constituents in groundwater.

Hydrogeochemical studies tend to be focused mainly on the occurrence and behaviour of the constituents which may cause deterioration of water quality, such as: nitrate, nitrite, ammonia, iron or manganese. Therefore, most recent papers and books concentrate mainly on these water components and only a small number of papers describe results of groundwater studies on the valuable water components as calcium or magnesium.

A large number of mineral water brands are present on the market, some of them are very popular in particular regions or countries, and together they possess a wide range of Ca and Mg concentrations. Consumers do not know which range of concentrations is beneficial for their health and which is not, nor is it easy for them to obtain concise information. Considering the high number of studies confirming the beneficial effects of these elements in drinking water supported by evidence from experimental and clinical studies, it is difficult to believe that most countries as well as the World Health Organisation have not recommended guideline levels for Ca and Mg in drinking waters or at least in mineral waters. These concerns inspired the preparation of this volume.

As calcium is the fifth and magnesium is the eighth most abundant element on Earth, and along with Na and K, are the major cations in groundwater, Ca and Mg play an important role in controlling the groundwater type. The main intention of this book is to emphasise the role of Ca and Mg in groundwaters as well as in bottled mineral waters in order to raise the profile of their importance and to promote relevant changes in the international regulations.

This volume is, therefore, dedicated to hydrogeologists and hydrochemists as well as the many professionals involved and interested in studies of human health and also to anybody who wants to know more about these two minerals, in shortage or in excess, and their role in water.

The structure of this book is very simple as it is divided into three parts:

1 Origin and occurrence of calcium and magnesium in groundwaters.
2 Significance of calcium and magnesium for human health.
3 Calcium and magnesium in mineral and therapeutic waters.

It is generally known that the chemical composition of groundwater is mainly determined by the composition of the rock it is abstracted from, but depending on the geochemical processes, similar types of rocks may lead to a range of chemical constituents in the groundwater.

In the **first part** of this volume the reader can learn about the origin and occurrence of Ca and Mg in groundwaters. Afelt explains the differences in Ca and Mg content in pore water in 'divide' and 'slope' parts of a loess sediment basin. Miessner and Ruede present a long term experiment with cut hard rock samples which elucidate the problem why dolomite dissolution is considerably slower than that of calcite and which process leads to an excess in Ca^{2+} compared to Mg^{2+}. Papic *et al.* describe the role of Mg/Ca ratio in hydrogeochemical processes and as a drinking water parameter. Rozkowski and Rozkowski show the impact of human activity on Ca and Mg concentrations in groundwaters. This chapter explains what causes an increase in the concentrations of Ca and Mg within the waters in vadose zone. Verbovsek answers the question if weathering of dolomite controls the major ion geochemistry of groundwaters and which are the most important processes in the karstic aquifers in Slovenia. Mezga and Urbanc explain why the majority of groundwaters in carbonate recharge areas are supersaturated with respect to calcite and dolomite. Sappa *et al.* explain how geochemical modelling techniques were employed to identify the main processes responsible for the evolution of spring waters from the carbonate aquifers. Dibal and Lar present the main sources of Ca and Mg in the crystalline aquifers as well as in the aquifers in the sedimentary basins in Northern Nigeria. Pluta and Grmela find the answer for the question whether mine waters discharged from the coal mines in the Upper Silesian Coal Basin enrich the waters of the Odra river by a concentrations exceeding the limits of Polish law or not.

Calcium and magnesium in drinking water have many beneficial effects on human health, however, at very high concentrations they can also have some negative impacts. Calcium, for example, may block the absorption of heavy metals in the body and is thought to lead to increase in bone mass and prevent certain types of cancer. On the other hand, very high concentrations may affect the absorption of other vital minerals in the body. Magnesium is an essential element in cardiac and vascular functions, but high contents in drinking water may have a laxative effect, particularly in the case of water with high concentrations of magnesium sulphates. More information about the role of Ca and Mg for human health is provided in the **second part** of this volume.

Kozisek considers if the extensive data on Ca and Mg concentrations in water and the knowledge acquired has had an effect on the regulatory field? A deficiency of Ca and Mg poses at least a comparable health risk as does exceeding the limit for some toxic substances, most of which are regulated and even the evidence of their toxicity is much less convincing than evidence of beneficial effect of Ca or Mg. The answer for the question whether the chronic magnesium deficiency combined with a reduction in its concentration in the myocardium are the results of insufficient supply of this mineral in the diet is described by Wojtaszek and Pieniak. Kousa *et al.* suggest that further studies are needed to prove – or disprove whether the low Mg concentration in drinking water is associated with the risk of coronary heart disease.

In the **last part** of the volume different aspects of calcium and magnesium in mineral and therapeutic waters are presented.

Razowska-Jaworek warns that the most important nutrient constituents of waters in 75% of bottled mineral waters sold in Europe are observed in extremely low concentrations. Taking into consideration Ca and Mg concentrations, these waters do not differ from drinking waters from public supply. Kielczawa explains which minerals have the largest influence on Ca and Mg content in medicinal waters. Sziwa *et al.* consider different compounds of Ca and Mg in natural waters used in therapeutic treatment or as food (bottled waters). They answer the question whether brines with a very high Ca and Mg concentration in the form of chlorides or sulphates may be considered only as therapeutic waters, or also to supplement calcium and magnesium intake. Nghargbu *et al.* explain why, although Ca and Mg in the West African springs are at a safe level for internal and external cures, they do not attract tourists there.

Most of the chapters collected in this volume are based on presentations from the international seminar: *Calcium and magnesium in groundwaters – distribution and significance* which was held in Katowice in 2012. I hope that this book will inspire the hydrogeologists to carry out further studies into Ca and Mg in groundwaters, and remind the medical profession of the importance of water as a source of Ca and Mg. The book should also bring the reader closer to the subject of Ca and Mg in drinking and mineral waters.

All the authors of the chapters of this book are interested in your experiences and questions on Ca and Mg in groundwaters and on the guidelines for Ca and Mg contents in drinking and mineral waters. I encourage you to send me any such observations via me at lidia.razowska-jaworek@pgi.gov.pl

I acknowledge my colleagues from the Upper Silesian Branch of the PGI-NRI for encouraging me in the process of editing this book and thank the support from Nick Robins from the IAH as well as Dr. Zofia Walencka from the Medical University of Silesia.

Lidia Razowska-Jaworek

Foreword

About 99% of the chemical make-up of pristine groundwater comprises the parameters HCO_3, Na, Ca, SO_4, Cl, NO_3, Mg, K and Si. The most common shallow groundwater type is Ca-HCO_3 dominated and Ca(Mg)-HCO_3 dominated. Rainwater, which is naturally acidic but weakly buffered, falls on the ground. Part runs off across the surface to streams and part is evaporated or transpired by vegetation back to the atmosphere. The remainder percolates underground. As the percolating water moves downwards under gravity towards the water table, the acid attacks the minerals present in the rock. The most reactive minerals, including particularly the carbonate mineral calcite, have the greatest impact on water chemistry, taking firstly Ca and HCO_3 into solution and buffering the acidity. These various processes controlling the dissolution of minerals by groundwater are explained by various authors in this book.

On reaching the water table the groundwater flows down the prevailing hydraulic gradient with residence times underground varying from just months to many thousands of years. Groundwater attains a pH that reflects the containing rock mass. This is most commonly buffered at around pH 6–8 but extremes can occur, while acid buffering is less easy to achieve in crystalline, non-carbonate bedrock conditions. Thus Ca is commonly dominant in groundwater and Mg may also be present in subordinate quantities wherever source rocks containing Mg, such as dolomite, have been in contact with the groundwater. A number of chapters in this book explore the relationship between lithology and the occurrence and distribution of Ca and Mg as well as the information to be gleaned by examining the Ca/Mg ratio.

The cations Ca and Mg are the key to the hydrochemical make up of all groundwaters. They are also critically important components of drinking water, satisfying essential mineral requirements of the human system that also protect it from illness. There are optimum ranges of concentrations in drinking water, particularly for Ca, with too little and too much both being detrimental to human health. This importance, however, contrasts starkly with the scarcity of regulation recommending such a range, and the general public are left in ignorance of any guideline levels towards good health. Although average Ca and Mg concentrations are printed on the label of bottled drinking waters in Europe and elsewhere, there is again no guideline information saying whether the stated concentrations are too high or too low for human wellbeing.

One of the reasons for lack of regulation is the issue of water hardness. Hardness has long been recognised as a scale of the effectiveness of soap to make a lather, and causes lime-scale build-up preventing harder waters from being used for boiler-feed water. Water supply undertakings ensure that drinking water supplies comply with guideline levels for hardness (where such guidelines exist) by blending harder and softer waters, but they make no attempt to inform the public about the hardness of the supply water.

The beneficial effects of Ca to the human system are described by various authors in this book, whereas the effects of Mg can be either positive or negative. Epsom Salts (magnesium sulphate) it would seem, dates back to the seventeenth century. The laxative effect was best exploited by the Europeans who enjoyed the new rural spa centres that emerged in the nineteenth century. The purgative properties of groundwater naturally containing Epsom Salts may have provided relief to some (and considerable discomfort to others).

This book has two key objectives. The first is to generate increased interest in the hydrochemistry of Ca and Mg as important cations in the make-up of groundwater quality. The second is to awaken interest, in those professionals working with drinking water quality, in the importance of Ca and Mg concentrations in groundwater and to the optimum range of concentrations that should be set for human wellbeing. The outcome from these objectives is renewed interest on the occurrence of Ca and Mg in groundwater and the setting of guideline ranges of concentrations for Ca and Mg by regulators worldwide. There is also need for explanatory information to be made available to the consumer.

This book is, therefore, a unique collection of papers that focus on an important area of hydrogeology that has not previously attracted a thematic set or volume. It is to be hoped that the work does indeed rekindle interest in Ca and Mg in groundwater and that it will generate some excitement towards setting guideline levels for drinking water.

Nick Robins
Editor-in-Chief

About the editor

 Lidia Razowska-Jaworek is hydrogeologist and hydrochemist at the Upper Silesian Branch of the Polish Geological Institute-National Research Institute in Sosnowiec, Poland. She graduated from the University of Silesia in Katowice, Poland and the University of Keele, UK.

She received an award for her doctoral thesis: "Hydro-geochemical changes in the Czestochowa region caused by the flooding of the iron mines" from the International Mine Water Association in the category: Hydrochemistry. She is engaged in a wide range of hydrogeological studies including: regional hydrogeology, aquifer pollution and protection, hydrogeochemical modeling, CO_2 storage in geological formations, acid mine waters and mineral waters and is also interested in medical geology. She is a regional editor of the Hydrogeological Map of Poland, author of 33 and co-author of 58 articles and co-author of 6 books and was an editor of the IAH SP volume "Nitrate in Groundwaters".

She was a chairman of the seminar "Calcium and magnesium in groundwater – distribution and significance" in Katowice in 2012, and is author of 3 and co-author of 5 articles on the importance of water chemistry and pollution for human health.

Part I

Origin and occurrence of calcium and magnesium in groundwaters

Chapter 1

Ca and Mg in loess pore moisture

Aneta Afelt

*Faculty of Geography and Regional Studies, University of Warsaw,
Warsaw, Poland*

ABSTRACT

Experimental field research was carried out to obtain pore water in three sites located within a
closed depression basin developed in loess on the top part of the Sandomierska Upland: water
divide, slope and in the bottom of the depression. At each experimental site the pore water was
sampled from the unsaturated zone, from depths of 1.5 and 4 m with the use of a system of
soil moisture samplers equipped with a teflon-quartz cube. The concentration of Ca and Mg
in the pore water was determined using the ICP MS method. The concentration of Ca in the
pore water is the highest and is stable over time: the highest values were found in the slope and
the watershed (approximately 500 mg/l on average), while concentrations for 'bottom' were
10-times lower and reached up to 50 mg/l. The Mg concentration is about 15% of Ca on the
slope and the water divide sites, however, it rises up to 20% in the 'bottom'. Variation in chemistry between the experimental sites is related to the pH of the pore water and the mineral components in the loess. Intensity of geochemical reactions is closely related to water circulation
within the loesses: intensive surface runoff and low effective porosity cause preferential accumulation of water in the 'bottom' which favours leaching of carbonate compounds ($CaCO_3$). In
addition a high concentration of Ca and Mg in all locations is a result of a strong porous water
binding effect by the rock matrix which in turn is a consequence of a recharge deficit.

1.1 INTRODUCTION

Loess is sedimentary rock of aeolian origin, polymineral, with a complex development
of components and sedimentation cycles. Pye (1995) points out that the shallow loess
layer comprises 10% of the global land area. There are three main groups of mineral
composition: quartz (80–85%), clay minerals (average 10%) and calcite (5–10%).
The ratio varies between different areas and can change between adjacent cycles of
sedimentation. Minerals in loess form a complex spatial structure of micro-aggregates
and aggregates (Grigorieva, 2001). The mineralogy of the loess and the water circulation conditions cause a variety of forms and intensity of geochemical processes. The
course of geochemical changes is regulated, among others, by its chemical content and
the types of water in the pores. Water movement in the loess pore space is controlled
mainly by capillary forces, and to a lesser extent, gravity, due to the small diameter
of the pores in the rock. As a result, there is no independent groundwater table in
loess and usually an extensive vadose zone. A low filtration coefficient for recharge
promotes surface retention and formation of hypodermal outflow.

The vadose zone is one of the most dynamic geochemical phases in the underground water circulation system. The process of porewater chemistry formation is complex and essentially depends on the characteristics of the individual minerals in the rock, on permeability and on supply conditions. The aim of the study was to investigate the seasonal variation of the concentration ions in the loess pore water, including calcium and magnesium as macro-elements.

The study of loess pore water was limited to the zone below the rhizosphere level. The hypothesis is tested that the changes that affect geochemical characteristics are determined only by moisture changes resulting from a sediment moisture tension gradient or gravity. Circulation of pore moisture was analysed as a function of input, precipitation recharge, and changes in concentration at the depth of 1.5 m and 4 m. The field experiments were conducted *in situ* for two seasons between April 2002 to April 2003 within the vadose zone of the upper young loess in the Sandomierska Upland (south east Poland).

1.2 MATERIALS AND METHODS

1.2.1 The experimental field

A genetically, age-synchronous and homogenous part of a loess aeration zone, within the land phase of the hydrological cycle, was selected for study. Influent seepage is the mechanism for transport in the pore spaces of the rock. The site is located in the central part of the Sandomierska Upland, interriverine-area of the Czyżówka and Opatówka rivers, in the vicinity of the Łukawa village (Figure 1.1). A number of closed depressions are the dominant features of the loess plateau relief. Topographically they form small blind drainages of various sizes. The blind drainage depressions, called in Polish 'wymok', are a specific type of loess deformation made by the infiltration process where the bottom of the depressions are periodically flooded by snowmelt or rainfall surface water retention. The study focusses on one of these depressions with sampling at depths of 1.5 and 4 m.

The study area is 8 km² and the stratigraphical and lithological uniformity was maintained during deposition under stable conditions. The site is isolated from anthropogenic activities including agricultural activities. The study has been conducted in three morphological positions representing experimental fields (Figure 1.2): 'water divide', 'slope' and the depression of the drainage basin, 'bottom'. Parts of a drainage basin that form a separate physical unit and have high internal homogeneity can be a blind drainage unit (Lechnio, 2008).

1.2.2 The mineral composition of loess

In order to identify the mineral composition of loess, rock samples were collected from the top zone (1.5 m below the ground) and bottom zone (4 m). The samples were subjected to quantitative and qualitative analysis by two methods: derivatography (SetarmLabsys TG-DTA12 camera) and X-ray diffraction (X-ray reflexive diffractometer DRON2.0).

For the derivatographic test, loess sub-samples, weighing 30 to 60 g, were used. They were separated from the host rock by rinsing 200–250 g of loess to prepare the

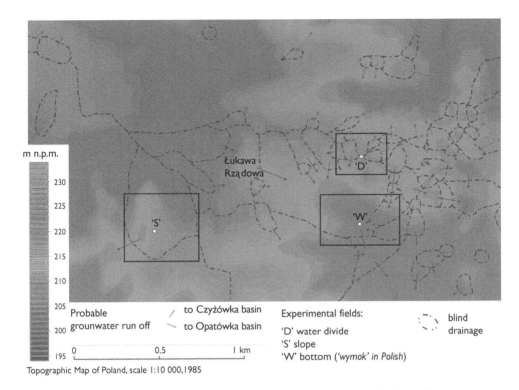

Figure 1.1 Field study location (Lukawa near to Sandomierz, S-E Poland).

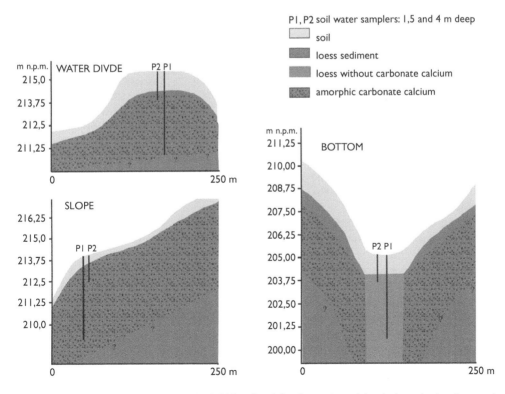

Figure 1.2 Experimental fields:'water divide','slope' and the depression of the drainage basin – 'bottom' (location see Figure 1.1).

Table 1.1 Specification of the quantitative contribution of minerals building loess, the upper young loess (Lukawa near Sandomierz).

Experimental field	'Water divide'		'Slope'		'Bottom'	
Depth [m]	1.5	4.0	1.5	4.0	1.5	4.0
Clay minerals	12.0–13.4	15.3–18.8	19–21.8	18.8–21.3	11.1–12.4	10.2–11.5
Beidellite	5.0–6.4	4.1–5.5	0.0 –2.2	0.0–1.1	1.2–1.6	0.9–1.4
Kaolinite	0.3–0.8	2.5–5.4	6.9–7.2	3.6–6.2	0.0–0.2	1.3–1.5
Illite	4.1–5.5	7.3–8.7	9.9–14.6	12.6–16.6	7.5–11.2	7.2–9.4
Goethite	1.3	1.7	2.0	2.8	2.4	1.4
Calcium carbonate (+ D – dolomite)	10.2 + D	8.3 + D	8.5 + D	8.1 + D	–	–
Quartz (+ potassium feldspar; plagioclase; mica)	75.1–76.5	71.8 –74.7	67.7–70.5	67.8–70.3	85.2–86.5	87.1–88.4

Minerals (%)

clay fraction for analysis (in dry and wet air) for each test point. Given the mineral composition of the multicomponent loess, X-ray analysis was carried out firstly for the non-clay minerals, and secondly an interpretation was then carried out to determine the types of clay minerals in the sub-samples. The radiographs were used to analyse three conditions: (1) the air-dry state, (2) after saturation with ether glycol, and (3) after ignition. (Table 1.1).

The analysis of the diffractograms confirmed the presence of clay minerals within the loess, distinguished in the deryvatographic study. The other minerals recognised are identified in Table 1.1 as 'other' and are quartz (predominant), plagioclases, micas and potassium feldspars. The analysis also revealed the presence of dolomite and siderite detected on testing sites 'water divide' and 'slope'. The tested loess consists of three clay minerals: kaolinite, illite and beidellite (Table 1.1). Grabowska-Olszewska (1989) classifies high hydrophilicity, plasticity, the ability to ion exchange, adsorption and swelling as the basic characteristics of this group of minerals. Kaolinite has two-layer packets, 1:1 and the inter-packet ion is potassium. Sorption of cations ranges from 1 to 15 mval/100 g capacity and it is the lowest amongst layer clay minerals. Exchangeable cations on the surface of the plates are located as follow: Fe, Ca, Mg, K, Na, Ti, Mn.

Kaolinite often forms spatial structures together with illite ($[Si_{6.9} Al_{1.1} O_{20}] (OH)_4$), which contains comparatively less potassium, and more water. It is the most common clay mineral found in loess and carbonate rocks. The structure of illite is low swelling. The cohesive force is weak, so their structure tends to be chaotic and easy to separate. In the octahedral layer, substitutions of Mg and Fe instead of Al can occur. Its sorption capacity is low, at around 20 to 50 mval/100 g.

On the other hand, beidellite ($A_{14} [Si_{7.33} Al_{0.67} O_{20}] (OH)_4$), a variety of montmorillonite, is built of a three-layer packet 1:2. Na ions are present in the interlamellar space, but Ca is also possible and both may be present. Interlamellar ions do not bind the structure, thus there is often organic matter and water present causing intense swelling, only slightly less than montmorillonite. The Al:Si ratio is 1:1. The Al tetrahedron is substituted by Si. However, substitution of Al by Fe (III) in the octahedra is also possible. The sorption capacity is about 80 to 120 mval/100 g.

Significant differences in the clay mineral content in the loess, especially in case of kaolinite and illite (Table 1.1), suggest a relation of these components with the feldspar weathering process. But it is not clear whether this is a result of geochemical changes, to which the primary minerals are subjected, or the result of a mechanical out-washing process (suffusion) removing the smallest particles within the filtration stream (Afelt, 2007). The differences in the clay mineral content in the vertical profiles can be used as an indicator of the mechanical suffosion rate. Therefore, additional to the normal geochemical processes, mechanical movement of minerals can influence the differentiation of secondary pore waters in the vertical profile.

Calcium carbonate, a mineral recognized as component of loess (Table 1.1), occurs in two forms, as a crystal or in amorphous form. Dwucet (1999) states that environmental conditions during the formation of the loess cover, were characterised by high carbonate contents. Hence their common presence as the original and secondary elements, resulting from the process of precipitatation of migrating solutions. Admixture of dolomite indicates the participation of these two carbonate forms.

Differences in the calcium carbonate content and its near absence in 'bottom', is the most important feature of the environment that is subjected to geochemical alteration. This mineral plays a key role in shaping the structure of loess; Ca is responsible for the aggregatisation of grains and particles. In saturated conditions where gas exchange is limited and inflow is accompanied by the decomposition of organic matter, acidic loess (pH 5.5) is formed. Calcium carbonate is then subjected to intense dissolution and removal of Ca (chemical suffosion process) resulting in peptisation of the sediment (decomposition of aggregates).

Goethite (α-FeOOH), which is a hydrated iron oxide, is commonly found in loess. Its presence in the precipitate of iron is the result of the transition to a sparingly soluble compound. The variable content of goethite in certain experimental fields and in the vertical profile as well as the admixture of siderite in the 'watershed' and 'slope' experimental fields (Table 1.1) is characteristic for geothite. The deficit of siderite in the 'bottom' profile is probably due to its complete dissolution, as well as of dolomite and carbonate in the geochemical conditions outlined above.

Quartz (SiO_2) is the predominant loess mineral and occurs in the form of grains. It is one of the most conservative minerals. Its solubility is limited, and in the hydrochemical cycle is considered to be practically insoluble (Hem, 1985). The weathering process of the quartz grain surface is visible in SEM (Scanning Electron Microscope) images in the form of a coating or crust. It indirectly determines the quality of the course of weathering conditions (Grabowska-Olszewska et al., 1984; Grabowska-Olszewska, 2001) (Figure 1.3). A bright, glassy, smooth amorphous coating indicates fast-paced ventilation (in 'bottom' only). However, when the silica was being precipitated at moderate speed the so-called 'inverted plates' and dull crust (typical in the case of quartz grains bonds of the bridge type) were formed in a slow process of weathering. In the experimental fields there are durable cementing bonds between the grains rather than the characteristic 'inverted plates' (Figure 1.3). Quartz grains are the main element of loess structure and forms of binding between them and other mineral elements control the stability of the rock structure (Figure 1.3).

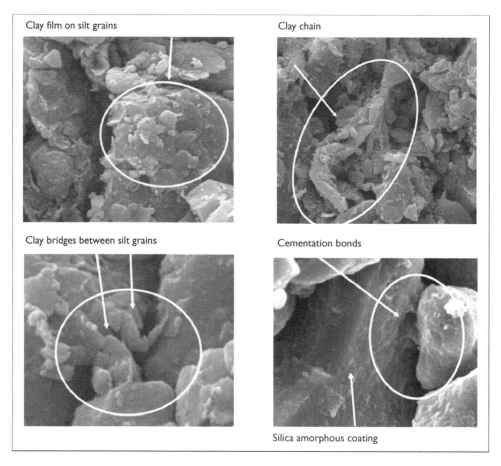

Figure 1.3 Microstructural surface of the upper young loess (Lukawa near Sandomierz); magnification 1000–500.

The analysis of the loess images under high magnification shows that the environment with high moisture content and slow gas exchange (with frequent water logging of sediments) is the most favorable condition for geochemical transformation of the loess sediment. Potassium feldspar (K [Al Si$_3$ O$_8$]), plagioclase, and micas occur in the loess in small amounts. They are typical examples of silicates (Si–O–Al). Hydrolytic decomposition of potassium feldspar leads to the formation of secondary minerals, mostly kaolinite or illite.

1.2.3   Method for obtaining pore solution

Soil water samplers were used to obtain repeated pore moisture in situ, while maintaining an undisturbed sediment structure. The samplers comprise three basic elements (Wilson, 1980), a porous cup, the solution receiver and vacuum pump. The principle is based on creating external negative pressure using the vacuum pump.

A closed, hermetic system tends to produce hydraulic relation between the rock medium and the porous cup. If the suction exceeds the physical energy retaining the pore water in the rock some of the water will be collected by the sampler.

The geochemical characteristics of loess in different experimental fields was best sampled with a Prenart quartz-teflon probe. It allows a low ionic determination threshold as the chemical stability of the porous cup is chemically stable and the sampling bottle is made of soda glass (Wilson, 1980; Afelt, 2003). The sampler (Prenart Equipment ApS 2004) accesses a pore diameter <2 μm, it has a porous surface of 33 cm^3, hydraulic conductivity is $3.31 \cdot 10^{-7}$ cm/s, maximum vacuum production cpability is 800 to 900 mb and the effectiveness of the vacuum is 6 ml/cm^2/24 h when the negative pressure value is 600 mb in the zone of full saturation. The coincidence in the pore diameter range of the sampler and tested pore medium was of particular importance for the experiment. This created an opportunity to establish hydraulic contact between the device and loess.

Each of the experimental fields has been equipped with two samplers, installed at 1.5 m below the surface) and 4 m below the surface. Sampling was carried out through two hydrological years between April 2002 and April 2004. The start of the sampling assesses the water system at the end of the winter. In order to maintain the constant, initial conditions for all samplers, a vacuum of 64 kPa was applied equivalent to a suction of approximately 2.4–2.6 pF. Pore water samples were collected in polyethylene cups, then stabilized with 1% nitric acid solution. The samples were stored in a refrigerator at −15°C. Time needed for pore water production varied depending on experimental fields, the range of recharge and season of the year.

1.2.4 The chemical composition of the pore solution

The main criteria for the selection of the analytical method were the varied volume and size of pore moisture samples (up to 0.5 cm^3) and Inductively Coupled Plasma Mass Spectrometry (ICPMS) was applied. ICPMS is recognised as one of the best methods of analysis of metal element concentrations in water (Becker & Dietze, 1998). An advantage of this method is the low determination threshold, high quality data, reliability and reproducibility of results, and the ability to analyse numerous ions and elements in a single sample.

1.3 RESULTS AND DISCUSSION

Loess pore water is subject to atmospheric recharge. The chemical composition of the pore water is formed by the sediment mineralogy and the intensity of the recharge. Local recharge conditions depend on conductivity and is characterised by a low filtration coefficient 0.1 to 0.03 m/d (data obtained from the quantitative analysis of the microstructure in SEM, Afelt, 2005), and this contributes to rapid surface runoff from the slopes. Under these conditions, a significant reduction in recharge takes place in the 'slope' profile and the 'bottom' is flooded with lateral inflow. As a result, in the 'water divide' and 'slope' profiles, the loess structure is unsaturated with a high possibility of gas exchange and in 'bottom' the pore structure is saturated and gas exchange is limited.

Table 1.2 The specification for the total concentration of tested elements in loess pore moisture, the upper young loess (Lukawa near Sandomierz).

| Solution concentration (mg/l) | Experimental field | | | | | |
| | 'Water divide' | | 'Slope' | | 'Bottom' | |
	1.5 m	4 m	1.5 m	4 m	1.5 m	4 m
Maximum	343.4	336.4	155.0	–	106.1	101.6
Medium	313.6	293.4	149.7	–	57.6	71.1
Minimum	264.8	220.2	143.3	–	12.3	57.9
Variation coefficient	0.3	0.4	0.1	–	1.6	0.6

Infiltration in porous loess is driven by gravity and the gradient of pore moisture tension in the rock. The concentration of elements in the pore moisture increases with depth. The total concentration of elements in the pore waters of the loess is summarised in Table 1.2. The largest concentration of elements in the pore water was found on the 'water divide', the smallest, some three times lower, was recorded in 'bottom'. The most stable mineralisation of pore water during the study period occurred in 'slope' which is related to the strong physical bond with the rock as the result of deficiency of recharge. The higher the moisture content and its variation in time, the greater the diversity of the total concentration of the pore water. This relationship is well illustrated by the variation in the total concentration of moisture in 'bottom' (Table 1.2). At a depth of 1.5 m below the surface, in periods of full saturation and flooding, mineralisation of is the lowest. With a subsequent decrease in moisture content, the concentration of elements in the moisture increases. And at 4 m below the surface, where the humidity is less variable in time, the variability of the total concentration of elements in solution is halved.

The variability in the concentration of elements in time is shown by candle chart (Figure 1.4). This type of chart is traditionally used in economics (Murphy, 1999). Its usefulness in the assessment of the concentration of substances in moisture is based on the clarity of data interpretation, and simple and clear visualisation of the variability in concentration over time. 'White candle' means an increase in the element concentration in moisture in time, 'black candle' a decrease in the concentration in moisture in time. The upper and lower marks are the minimum and the maximum concentration during the experiment. Input data are the value of element concentrations in subsequent periods treated as a set of data arranged from the beginning to the end of the experiment. Another advantage of this method of data presentation is the visualization of the results regardless of the number of samples for each experimental field.

The process of dissolution of elements is strongly associated with their properties and acid base reaction of rock. Within the experimental fields representing an alkaline environment (alkaline pore water) (pH 7.5, 'water divide' and 'slope') the concentration of particular elements in the pore water is similar. On the 'slope', apart from Si and to a lesser extent Ca, the concentration of the other elements is lower (Figure 1.4). The highest concentrations (<1 mg/l) are Ca, Mg, Si, K and Fe.

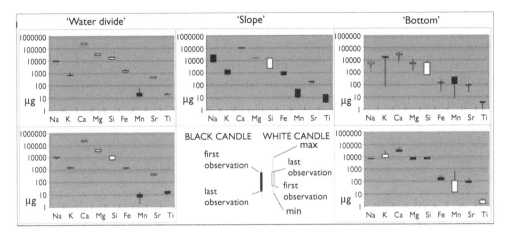

Figure 1.4 Variability of the ion concentration in loess pore moisture, the upper young loess (Lukawa near Sandomierz).

Table 1.3 Ca and Mg concentration in loess pore moisture, the upper young loess (Lukawa near Sandomierz).

Ion	Experimental field	Concentration in pore water (mg/l)			Variation coef.	% in total concentration	Mg/Ca (min–max)
		Average	Min	Max			
Ca	'bottom' 1.5 m	23.1	7.7	44.6	1.6	40.2	0.21 (0.21–0.24)
	'bottom' 4 m	35.7	27.3	49.8	0.6	50.2	0.22 (0.21–0.17)
	'slope' 1.5 m	107.6	99.9	112.9	0.1	71.9	0.15 (0.16–0.15)
	'water divide' 1.5 m	248.8	210.6	272.6	0.2	79.3	0.13 (0.13–0.14)
	'water divide' 4 m	232.4	177.3	264.4	0.4	79.2	0.16 (0.14–0.17)
Mg	'bottom' 1.5 m	4.7	1.6	10.7	1.9	8.2	–
	'bottom' 4 m	7.7	5.8	8.6	0.4	10.8	–
	'slope' 1.5 m	16.4	15.8	17.3	0.1	11.0	–
	'water divide' 1.5 m	33.2	28.4	37.1	0.3	10.6	–
	'water divide' 4 m	36.5	24.9	44.1	0.5	12.4	–

Changes in concentrations of element over time on the 'water divide' are similar at both 1.5 and 4.0 m depth. The greatest variation in concentration takes place in 'bottom' (Figure 1.4), representing an acidic environment (pH 5.5). A large variation in element concentrations in both the upper and lower part of the profile occurs, in particular Mg, Ca, K and titanium.

Ca is an extremely dominant element in the pore moisture. It comprises almost 80% in the 'slope' and 'water divide', decreasing to 40–50% within 'bottom' (Table 1.3). Its ionic radius (0.99 Å) binds it by substitution with Na and the lanthanide chain (Hem, 1985). As a result, under high concentration conditions (for example, under a recharge deficit), Ca is removed by Na. This process is of importance to the structure of loess. Ca leaching (chemical suffusion) weakens the structural bonds between mineral elements in the rock, and in the extreme conditions of Ca deficit it can lead to disequilibrium in the stability of phase bonds. In the process of chemical weathering nearly all the Ca is in the form of bicarbonate and is soluble in water and gradually leached from the rock (Hem, 1985). The lack of calcium carbonate and dolomite in the mineral composition of 'bottom' is significant (Table 1.1). These components were completely removed in the dissolution process and discharged elsewhere. The geochemical response to the calcium carbonate and dolomite presence in the profile of 'bottom' is an increase in the strontium concentration in the pore moisture, strontium substitutes calcium in carbonates, but as an element with lower solubility it is more resistant to the process of hydration which with the low permeability of loess makes drainage of the loess more difficult (Afelt, 2007).

Mg is the second important element in respect to the content of the porous moisture in 'water divide' and 'slope' (Figure 1.4). It contributes 10 to 12% of the total mineralisation. Within 'bottom' its concentration is up to five times lower, with only a slight reduction of the total percentage which is 8 to 10% (Table 1.3). Constant ion valency and its construction close to the one of noble gas determines its behaviour in simple crystal structures (Hem, 1985). Its small ionic radius (0.66 Å) binds it in close relationship with substitution for Fe, and both constitute common chains of isomorphic mixed iron and magnesium crystals. Mg coexists with Fe in illite and takes part in substitution of Al in octahedrons. Common forms of Mg-Ca doubles salt are formed (no substitution due to significant difference in ionic radius).

In hypergenic conditions Mg ions become mobile (the most important minerals are dolomite, magnesite and chlorite). In the presence of chloride and sulfate (VI) ions, Mg compounds change into soluble chlorides and sulphates during weathering and are leached from the rock (Hem, 1985). Adsorption by clay minerals is less than that by Ca. Dolomite, present in the mineral composition of sediment in the water divide' and 'slope', is probably the main source of Mg in the loess. The mineral deficit in the field of 'bottom' is probably related to the chemical weathering process and to its drying out during the sampling period.

The quantitative ratio of Mg to Ca is variable depending on the location (Table 1.3). Despite the significantly lower levels of Mg in the pore waters of 'bottom', it is here where Mg concentration in relation to Ca is the highest (0.2). This stems from a decrease of the Ca as a consequence of leaching of carbonates. Czarnecki & Sonceva (1992) show that decalcification in 'bottom' decreases at least to a depth of 10 m (the depth of the borehole in neighboring 'bottom'). The water recharging 'bottom' from rainfall and runoff is the other source of both Ca and Mg. Granulometric analysis shows that the 'bottom' serves also as storage area for particulate clay minerals, predominantly illite (Afelt, 2005, 2007). The process of mechanical suffusion or outwashing of allochthonous clay minerals within filtration stream, accounts for part of the increase in the concentration of illite in the pore space of 'bottom' (Table 1.1). Illite may have originated from weathering of potassium feldspar in the mineral composition of the rocks, with an increase in the concentration of K and Si (Table 1.1, Figure 1.4).

A significant increase in the concentration of Na ions may also play a role in the disruption of the Ca-Mg relationship (Figure 1.4) as well as removing a ions in solution (Hem 1985). The analysis of the loess pore space structure shows that the highest porosity is present in the experimental field of 'bottom' at 57% (Afelt, 2007), while moisture content fluctuations throughout the year are insignificant and tend to be fully saturated.

The impact on the differences in the concentration of elements in the loess pore water (mainly differences in Ca and Mg), is created by the conditions in the source of the moisture. Due to the deficiency of pore water within the experimental fields 'water divide' and 'slope', applied external pressure of 2.4–2.6 pF allows the capillary water structure to recover. This means that the ratio of soluble component concentrations in the water is high. The sample volumes are small compared to the applied pressure. The high saturation of loess structures in experimental field 'bottom' allows efficient recovery of pore moisture. However, although it is capillary water, it is poorly bound with the rock and has a lower amount of solute.

1.4 CONCLUSIONS

Differences in the concentration of Ca, Mg as well as coexisting Sr, Na, K, Ti, Si, Fe, Mn in the pore water moisture in the vadose zone of loess have been found. Variability reveals a relationship to the local relief of the surface as a consequence of the water circulation conditions and the low filtration coefficient of the rock. Atmospheric recharge in blind drainage 'bottom' is preferred as it receives additional recharge from surface runoff and from interflow from the slopes.

Sampling of pore waters is dependent on the moisture content in the sediment. The movement of water in the pore space is capillary, and water saturation of loess is full or close to full only in the vicinity of the sediment column beneath the 'bottom' (4.0 m depth). In other experimental fields only periodically increased moisture content conditions occur. An externally generated negative pressure of 2.4–2.6 pF in 'water divide' and 'slope' sites enabled pore moisture to be sampled despite it being physically bound with rock. Pore water sampling in the experimental site 'bottom' did not pose any technical problems and was less physically bound to the rock.

Recharge and gas exchange differences determine the geochemical environments. 'Water divide' and 'slope' sites are alkaline environments for pore waters, with high concentrations of Ca (up to 80% of total mineralisation), and Mg as the subordinate component (10–12%). The mineral composition consists of calcium carbonate and traces of dolomite. The moisture in the loess during drainage periods at 'bottom') is acidic, decalcified, with limited atmospheric gas exchange. There is a typical reduction in Ca concentration in the moisture (40–50% of total mineralisation). Despite the lower concentration, Mg increases by up to 20%, along with Na, K and Si and a large increase in K (>100%) due to the weathering of potassium feldspars.

The Ca and Mg concentrations in loess pore waters need to be interpreted carefully according to the local physical and chemical environment.

ACKNOWLEDGEMENTS

The study, conducted in 2003–2005, was funded by KBN (State Committee of Science Research); research project 3P 04 E 007 23.

REFERENCES

Afelt A. (2003) An influence of soil water sampler substance on chemical composition of porous moisture, *Acta Universitatis Carolinae, Geographica*, 1: 5–20.

Afelt A. (2005) Infiltration Transformation of the Loess Structure, PhD Dissertation, manuscript (in Polish).

Afelt A. (2007) Influence of relief and water supply on the properties of pore water in loesses, Annales UMCS, 62: 223–242 (in Polish).

Becker J.S, Dietze H.-J. (1998) Inorganic mass analysis by mass spectrometry, *Spectrochimica Acta Part B*, 53(11), 1475–1506.

Czarnecki R. & Solnceva N.P. (1992) The small suffusive depressions near Sandomierz (part II), *Przegl. Geograficzny,* 104(1–2), 143–149 (in Polish).

Dwucet K. (1999) Lithogenesis of Vistulian Upper loess in Poland Uplands and Silesian Lowlands, *Prace Naukowe Uniwersytetu Slaskiego*, Katowice 1792: 1–163 (in Polish).

Grabowska-Olszewska B. (1989) Skeletal microstructure of loess – its significance for engineering-geological and geotechnical studies, *Applied Clay Science*, 4(4), 327–336.

Grabowska-Olszewska B. (2001) Collapsibillity of Polish Loesses in Unsaturated Zone, *Collapsic Soils Communique*, 2, 4–9.

Grabowska-Olszewska B., Osipow W. & Sokołow W. (1984) *Atlas of the microstructure of clay soils*, PWN, Warszawa.

Grigoriewa I.J. (2001) *Mikrostrojenie liossowych porod*, Nauka Interperiodyka, Moscov (in Russian).

Hem J.D. (1985) *Study and Interpretation of the Chemical Characteristics of Natural Water*, US Geological Survey Water-Supply Paper 2254.

Lechnio J. (2008) Modeling of the selected aspects of geosystems functioning, *Problemy Ekologii Krajobrazu*, 20, 184–204 (in Polish).

Murphy J.J. (1999) *Technical Analysis for the Financial Markets*, New York Inst. of Finance, New York.

Prenart equipment ApS (2004) Prenart, Frederiksberg.

Pye K. (1995) The Nature, Origin and Accumulation of Loess, *Quaternary Science Review*, 14, 653–667.

Wilson L.G. (1980) *Monitoring the Vadose Zone: A Review of Technical Elements and Methods*, United States Environmental Protection Agency, Environmental Monitoring System Laboratory, Las Vegas.

Chapter 2

The solubility of different carbonate rocks in natural and anthropogenically influenced waters

Paul Miessner & Thomas R. Rüde
Institute of Hydrogeology, RWTH Aachen University, Aachen, Germany

ABSTRACT

In carbonate areas, the dissolution of rocks is the main process that determines the hydrochemical character of groundwater, leading especially to high concentrations of Ca^{2+}, Mg^{2+} and HCO_3^-. Extensive karstification causes considerable landscape modification (e.g. ground breaks, sinkholes) even being harmful to humans if karst features appear in urban areas. This chapter focuses on the question if and how anthropogenic influences on (ground) waters affect their ability to dissolve carbonate minerals and rocks, by either accelerating or retarding the reaction. Short- and long term dissolution experiments have been used to test the solubility of two almost pure carbonate rocks in different water types. Results indicate slight differences in the ability of the different water types to dissolve carbonate minerals although exact effects and determining factors are not easy to quantify.

2.1 INTRODUCTION

Carbonate rocks cover approximately 10 to 15% of the earth's continental surface, and as they act as karstic aquifers provide potable and useable water to up to a quarter of the total population (Ford & Williams, 2007). Huge areas depend on these resources, making them essential for human health and development, for instance in the Mediterranean region. Two areas were chosen, where almost pure carbonate rocks represent the main aquifer in their respective distribution area: Jurassic dolomite from Sefrou, Morocco; and Devonian calcitic limestone from Wuppertal, Germany.

In Morocco rock and water samples were taken from the town of Sefrou, approximately 25 km south east of Fès, at the foothills of the Middle Atlas, where the Liassic karst aquifer of the Atlas Mountains is slightly overlain by Plio-Quaternary sedimentary rocks of the Saiss basin (Figure 2.1). Karstic aquifers hold up to 70% of the groundwater available in Morocco and the Liassic dolomitic limestone is a typical lithology in the High and Middle Atlas. It covers about 10% of non-Saharan Morocco (ca. 30 000 km²) and the water is generally of good quality with low mineralisation. Around Sefrou the dolomites dip gently to the north west and groundwater hydraulics change from unconfined in the Middle Atlas to confined underneath the Saiss basin. Recharge takes place through meteoric water inside the mountains, but due to overexploitation and a continuing precipitation deficit groundwater levels have been dropping since the 1980s (Amraoui *et al.*, 2005).

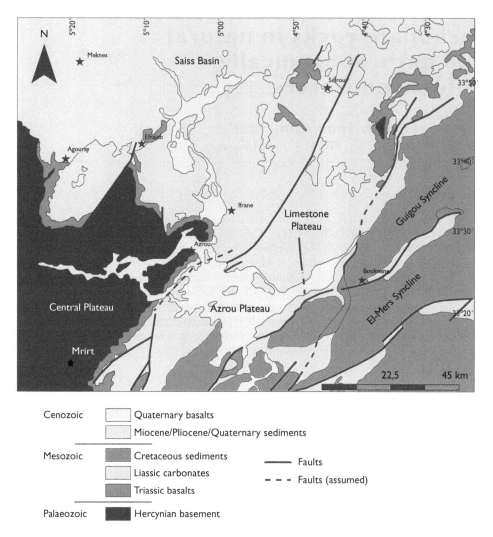

Figure 2.1 Geological map of the Tabular Middle Atlas, Morocco (after Hamidi *et al.*, 1997).

Close to the town of Wuppertal in western Germany rock and water samples were collected in an abandoned quarry, where almost pure Devonian limestones were previously won for cement production. The area is part of the northern Rhenish Massif, where Middle (and Upper) Devonian carbonates are surrounded by Lower Devonian clastic sediments (Figure 2.2). The so called 'Massenkalk' (compact and massive limestone) is up to 350 m thick and was deposited as huge reef complexes in the Rhenohercynian shallow sea. Due to a very low primary porosity and matrix permeability groundwater flow is restricted to fissured and karstified areas of the rock; a common phenomenon in this region (Kohlhaas, 1972). For ten hydrological years between 2000 and 2010 the average precipitation was about 1145 mm, with an evapotranspiration of 48% or 550 mm (Turc) respectively.

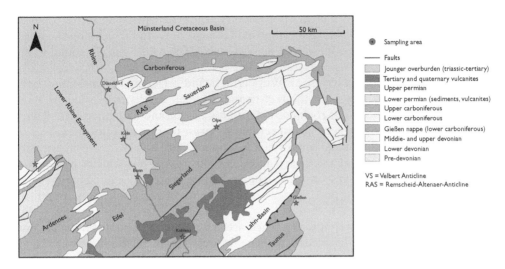

Figure 2.2 Simplified geological map of the northern Rhenish Massif including sampling area (after Henningsen & Katzung, 2002).

In the last 50 years numerous papers and books have been published that deal with carbonate dissolution in water and karstification in the laboratory and field scale, and the basic processes are well known (e.g. Liu & Dreybrodt, 1997). Due to the generation of carbonic acid and the lowering of pH the most important factor is the amount and availability of CO_2; but other parameters such as temperature, surface volume, surface structure, saturation state and flow velocity can have a huge influence on dissolution kinetics as well. In laboratory experiments dolomites generally show considerably slower dissolution rates than pure calcite, but because of the time factor in natural systems they are able to karstify equally well (e.g. Herman & White, 1985). The dissolution of calcite is also known to be a reversible process by the precipitation of the mineral regardless of scale (e.g. formation of travertine), a phenomenon that has not been reported for dolomite so far.

In spite of this knowledge the influence of other processes, especially through anthropogenically released ions, is not well known. Few papers deal with the sorption of metal cations on calcite (e.g. Zachara *et al.*, 1991). This process seems to hinder the dissolution by the formation of coatings or solid solutions; but hardly anything is published on other chemical influences. This issue might, however, be of interest, because karstic aquifers in urban areas are very vulnerable to the entry of various chemicals, e.g. by direct or diffuse input of sewage.

2.2 MATERIALS AND METHODS

The rock samples were taken as representative specimens of 5.6 kg (Sefrou) and 16.2 kg (Wuppertal) respectively from unweathered parts of the outcrops.

The Moroccan dolomite has a light grey colour and a very dense and massive matrix. Only a few microfissures are visible inside the rock with small and unconnected dissolution voids. The mineral density is around 2.8 gm/cm³ and the total porosity in the range of 0.8 to 1.5% by mercury porosimetry. The geochemical and XRD analysis (Rietveld) are shown in Table 2.1 and give an almost pure and stoichiometric dolomite with minor quantities of Cr, Mn, Sr, V, Cu, Ni, Zn, As, Br and Co found in the range of 2–50 ppm.

In the abandoned quarry near Wuppertal Middle Devonian limestones have three different lithologies from the reef to the forereef. Rocks from the core have been used for this study. The rock has a dark grey colour and a massive matrix with some calcitic veins that represent old refilled fissures; no open cracks are visible to the naked eye. It has a mineral density of 2.7 gm/cm³ and a very low total porosity of about 1%. Geochemical results in Table 2.2 show almost pure calcite with minor quantities of MgO and SiO_2. Cr, U, V, Cu, Ni, Zn, As, Y and La can be measured in very small amounts (<10 ppm), whereas Ba (80 ppm) and Sr (144 ppm) are a little higher.

Two water samples were taken in each area to represent the natural system and one anthropogenically influenced water type respectively. In Sefrou these were taken as bailed samples from Oued Aggai, a stream running right through the city centre; the natural sample near the cascades before entering town, and the 'anthropogenic' one in the middle of the medina (city), where you can recognise the direct input of sewage and dumping of solid waste along the stream. In Wuppertal the natural water sample is from an open groundwater lake in the same quarry that the rock sample was colleted. The other is untreated sewage from a wastewater treatment plant (Buchenhofen), which is approximately 4 km south east of the quarry and receives the major part of all domestic sewage from the town of Wuppertal. All physicochemical properties were measured in situ, whereas the inorganic hydrochemical content was analysed in the laboratory (Table 2.3). Because the volume of each water sample

Table 2.1 Geochemical and mineralogical properties for the Moroccan dolomite.

	Weight [%]	Mineral	Weight [%]
CaO	25.05	$CaMg(CO_3)_2$	98.2
MgO	18.90	$CaCO_3$	1.6
Fe_2O_3	0.51	SiO_2	0.2
Al_2O_3	0.34		
SO_3	0.27		
Na_2O	0.13		
K_2O	0.10		
P_2O_5	0.01		

Table 2.2 Geochemical properties for the Devonian reef limestone.

	Weight [%]		Weight [%]
CaO	54.61	Mn_2O_3	0.04
MgO	0.60	K_2O	0.03
SiO_2	0.51	S	0.01
Al_2O_3	0.14		
Fe_2O_3	0.10	Loss of ignition	43.61

Table 2.3 Hydrochemical properties for the water samples.

	T [°C]	pH	EC [μS cm⁻¹]	TDS [mg L⁻¹]	Ca/Mg ratio
Sefrou – Cascade	18.7	7.6	487	456.5	0.86
Sefrou – Medina	18.7	7.7	536	494.0	0.99
Wuppertal – Groundwater	16.0	8.5	362	264.0	1.80
Wuppertal – Sewage	22.3	6.7	865	553.7	2.61

	Ca^{2+} [mg L⁻¹]	Mg^{2+} [mg L⁻¹]	Na^+ [mg L⁻¹]	K^+ [mg L⁻¹]	Sr^{2+} [mg L⁻¹]	NH_4^+ [mg L⁻¹]	HCO_3^- [mg L⁻¹]	Cl^- [mg L⁻¹]	SO_4^{2-} [mg L⁻¹]	PO_4^{3-} [mg L⁻¹]	NO_3^- [mg L⁻¹]
Sefrou – Cascade	52.2	37.0	5.8	0.9	0.3	<0.01	329.5	<2.5	<5	<0.05	32.8
Sefrou – Medina	53.2	32.7	10.7	1.9	0.4	<0.01	363.1	5.6	6.0	0.8	20.4
Wuppertal – Groundwater	44.5	15.0	14.4	2.2	0.1	<0.01	128.1	11.5	40.9	<0.2	3.2
Wuppertal – Sewage	53.8	12.5	52.7	13.6	0.1	33.9	323.4	41.1	21.6	4.9	<0.9

	B [μg L⁻¹]	Co [μg L⁻¹]	Ni [μg L⁻¹]	Al [μg L⁻¹]	Cu [μg L⁻¹]	Zn [μg L⁻¹]	As [μg L⁻¹]	Cd [μg L⁻¹]	Fe [μg L⁻¹]	Mn [μg L⁻¹]	Pb [μg L⁻¹]
Sefrou – Cascade	12.4	40.0	12.3	4.1	<0.3	<0.7	0.1	<0.3	<10	<0.5	<0.3
Sefrou – Medina	27.7	40.0	19.0	7.0	1.1	2.9	0.3	<0.3	<10	<0.5	<0.3
Wuppertal – Groundwater	17.1	0.1	2.4	0.3	0.5	<10	0.2	<0.01	<10	120.0	<0.05
Wuppertal – Sewage	203.8	67.4	156.3	962.5	283.0	10.0	0.8	37.6	230.0	210.0	208.9

was not large enough to fulfil all dissolution experiments, similar waters had to be synthesised in the laboratory by adding standard solutions and salts to deionized water.

With these rocks and synthesised water samples two different sets of dissolution experiments were carried out; one to recognize short term reactions on a large surface and a second for long term effects on small surfaces, simulating natural groundwater systems.

The first set was designed as batch tests with a high water rock ratio of 25:1; 20 gm of finely crushed rock samples were shaken overhead with 500 ml of the different water types for 48 hours. For the long term experiments drilled cores of the Moroccan Liassic dolomite with a diameter of 38 mm and a length of 80 mm were used. The different waters were then forced through these hard rock samples inside a pressure cell for about 6 months. Because of the very low hydraulic conductivity of the rock matrix (K value of 4.5×10^{-12} m/s; hydraulic permeability of 4.6×10^{-19} m^2) transport is along joints and fractures. So, the solid cores were cut into two halves along the long axis and resampled to produce a well-defined artificial fissure (Figure 2.3).

In all experiments specific Electric Conductivity (EC), pH values and concentrations of Ca^{2+}, (Mg^{2+}) and HCO$_3^-$ were measured before and after the tests to measure dissolution effects and calculate saturation indexes and maximum solubility for the different minerals by the PC-code PhreeqC.

Figure 2.3 Experimental setup of the pressure cells and detailed view of a cut rock sample.

2.3   RESULTS

2.3.1   Bath tests

All batch tests with the different rock and water samples were done in duplicate, and results are shown in Figures 2.4 and 2.5. Natural water types are named 'Cascade' (Morocco) and 'Quarry' (Wuppertal) whereas the anthropogenic types are 'Medina' and 'Sewage' respectively.

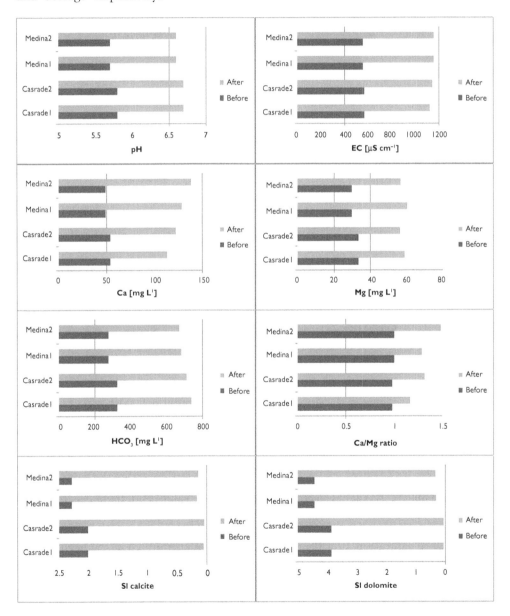

Figure 2.4 Results of the batch tests with Moroccan dolomite and water types.

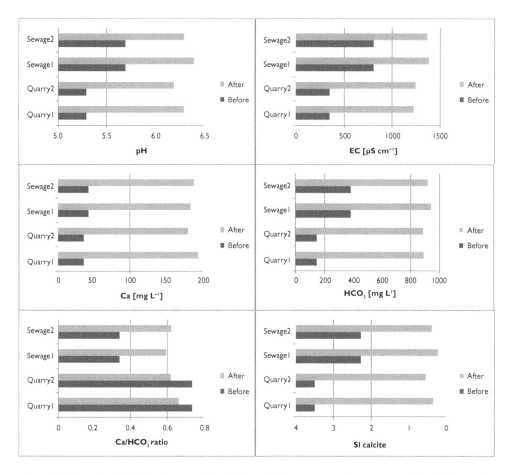

Figure 2.5 Results of the batch tests with calcitic reef limestones and water types from Wuppertal.

In the experiments with Moroccan dolomite all solutions start with EC values about 560 µS/cm and pH 5.7–5.8, which rise to 1100–1200 µS/cm and pH 6.6–6.7 respectively after shaking with the finely crushed rock sample for 48 hours. Because of the mineral dissolution during this time, there is an evident rise of the analysed ions: Ca^{2+} from about 50 mg/l to a maximum of 138 mg/l, Mg^{2+} from around 30 mg/l to 56–60 mg/l and HCO_3^- from about 270 mg/l to 680 mg/l (Medina) and 320 mg/l to 710 to 740 mg/l (Cascade) respectively. Most of these values and, therefore, total and relative rises are a little higher in the Medina water type. In addition all Ca/Mg ratios rise from values of 1 to 1.2 or 1.5 and saturation indexes for calcite and dolomite come close to saturation after being considerably undersaturated at the beginning of the experiments.

The second set with Devonian reef carbonate shows some similar results, but because of the larger difference in the two waters there are bigger variations. EC rises from 350 µS/cm to 1230 to 1250 µS/cm (Quarry) or 800 µS/cm to almost 1400 µS/cm (Sewage) respectively and pH values from 5.3 to 5.7 to 6.2 to 6.4. Ca^{2+} concentrations are about 40 mg/l before shaking and go up to nearly 200 mg/l afterwards, whereas HCO_3^- ends up between 880 and 940 mg/l despite the varying starting concentrations. All resulting solutions are slightly undersaturated with respect to calcite after showing initial values of −2.27 and −3.49; but this time the bigger changes appear mostly in the natural water type.

2.3.2 Pressure cell experiments

Figures 2.6 and 2.7 show the results of the two different pressure cell experiments with Moroccan dolomite, the natural cascade water in cell 1 and the anthropogenic medina water in cell 2. Input values of both reservoirs have been sampled and illustrated as well for comparison.

The first experiment (cell 1, Figure 2.6) was done for a little more than six months (4600 h) from January to July 2012. Measured pH values vary between 7.8 and 8.2 and are, therefore, always well above the reservoir values of 6.3–7.2. The EC of the outflow initially stays in the range of the input between about 520–540 μS/cm, before

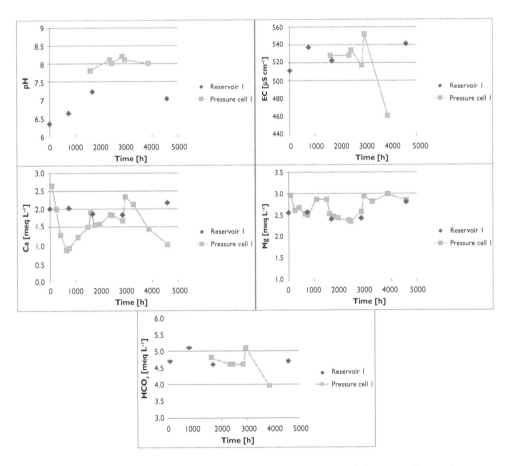

Figure 2.6 Results of the pressure cell experiment 1 with Moroccan dolomite and 'natural' water.

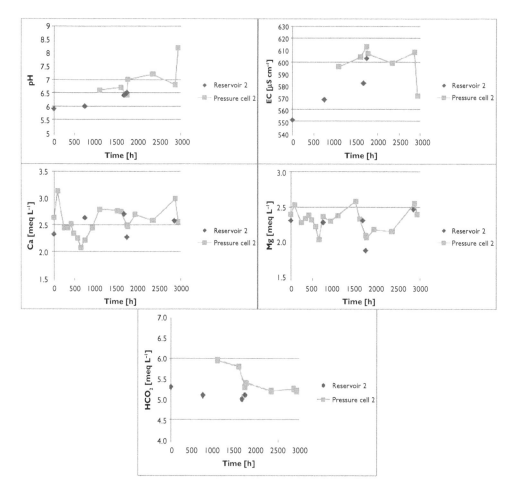

Figure 2.7 Results of the pressure cell experiment 2 with Moroccan dolomite and 'anthropogenic' water.

reaching the maximum value of 552 µS/cm at 3000 hours and dropping rapidly to 460 µS/cm afterwards. Ca^{2+} and Mg^{2+} values for the input water are more or less constant throughout the experimental time at about 2 meq/l for Ca^{2+} and 2.5 meq/l for Mg^{2+} respectively. In the outflow measured Ca^{2+} concentrations vary to a greater extent. In the beginning there is a rapid drop to the total minimum of 0.84 meq/l at about 650 hours, followed by a constant rise to 2.33 meq/l at 3000 hours and another rapid decrease to a final value of 0.99 meq/l. For Mg^{2+} changes are not so distinct and all values vary between 2.36 and 2.99 meq/l with a progression that is similar to that of Ca^{2+}. Most of the HCO_3^- concentrations in input and outflow are within the range of 4.6 to 5.1 meq/l but there is a noticeable drop to 3.95 meq/l between 3000 and 3800 hours of experimental time.

Cell 2 could only be operated from January to May 2012 (almost 3000 hours), because there was no adequate outflow for sampling and measuring after that time.

The fissure was probably mechanically blocked. The pH values in the input reservoir rise throughout experimental time from 5.9 to 6.5, whereas outflow values are generally above that and increase even more. They are between 6.4 and 7.2 up to 2860 hours and show an maximum of 8.2 at the end of the experiment. Specific electrical conductivity also rises in the input from 550 to 600 μS/cm, but again outflow values generally stay above that (596–613 μS/cm) up to 2860 hours. Contrary to pH the EC drops to the minimum value of 571 μS/cm at the end of the experiment. All input values for Ca^{2+} and Mg^{2+} are in the range of 1.9 to 2.7 meq/l but vary a little more than in cell 1. Their concentrations in the outflow of cell 2 evolve in the same way over the whole experimental time (Figure 2.7). Both show a first peak shortly after initiation and drop constantly afterwards to minimum values of 2.07 meq/l for Ca^{2+} and 2.04 meq/l for Mg^{2+} respectively at 650 hours. Ca^{2+} concentrations then rise sharply and stay in the range of 2.43 to 2.99 meq/l with minor variability until the end of the experiment. Mg^{2+} values also rise to the total maximum of 2.58 meq/l at 1500 hours, before quickly dropping to 2.07 meq/l at 1750 hours and finally rising to around 2.5 meq/l. The bicarbonate concentration in the input is almost stable with values between 5.0 and 5.3 meq/l. In the outflow concentrations are consequently higher but drop constantly from a maximum of 5.95 meq/l at 1085 hours to 5.2 meq/l at the end of the experiment.

2.4 DISCUSSION

The dissolution (and precipitation) of carbonate minerals in water is a widely recognised and an observed process in natural and laboratory systems. Although in reality it is a multi-step reaction through the genesis and dissociation of carbonic acid it can be simplified and formulated as:

$$CaCO_3 + H_2CO_3 \leftrightarrow Ca^{2+} + 2HCO_3^-$$

and

$$CaMg(CO_3)_2 + 2H_2CO_3 \leftrightarrow Ca^{2+} + Mg^{2+} + 4HCO_3^- \text{ (congruent)}$$
$$CaMg(CO_3)_2 + Ca^{2+} \leftrightarrow 2CaCO_3 + Mg^{2+} \text{ (incongruent)}$$

for calcite and dolomite respectively (e.g. Appelo & Postma, 2005). As carbonic acid is made from water and CO_2 these reactions are mainly favoured through the input of gaseous CO_2 and cause higher pH values. In addition they obviously bring Ca^{2+}, Mg^{2+} and HCO_3^- into solution and thus determine hydrochemical characteristics of many groundwaters. The reverse of this process is only known for calcite so far and is usually caused by the removal of CO_2 from solution e.g. through biological activity or increased turbidity. In the short and long term experiments these effects can be observed and partly quantified.

In the first set of batch tests with Moroccan dolomite and different water types the differences in the results are small but noticeable. As the almost doubling of EC and the rise of pH values reveal, dissolution of the dolomite has apparently taken place. The release of ions is in a comparable range in all tests, but there is a slight excess in relative and total values obtained for the cations in the medina waters, approx. 20 mg/l for Ca^{2+} and 5 mg/l for Mg^{2+}, whereas HCO_3^- shows no differences. Unfortunately it cannot be proven if this is due to the small amounts of heavy metals added to the solution or to the slightly lower pH value (5.7, compared to 5.8). Calculated masses of dissolved pure dolomite range from 172 to 191 mg/l for the cascade and 201 to 230 mg/l for the medina, and there is always an excess in Ca^{2+} between 0.43 and 1.1 meq/l. This proves the incongruent dissolution of dolomite with Ca^{2+} release, a process that can also be recognised in elevated Ca/Mg ratios of between 1.2 and 1.5 after the experiment. As the input waters have ratios of 1, this process will not take place in the natural system where congruent dolomite dissolution is assumed. All saturation indexes for calcite and dolomite are close to 0 (representing saturation) at the end of the batch tests; the dissolution process seems to be a fast reaction under the control conditions (large reaction surface) and finished after 48 hours of experimental time. Again the medina waters are less saturated with both minerals, but this could also be caused in the slightly lower pH value.

In the second set with Devonian calcitic limestone and water types from the area of Wuppertal, the dissolution of calcite is also apparent. The pH values show a rise of up to one unit from 5.3 to 6.3 in the natural water types and from 5.7 to 6.4 in the sewage, EC rises strongly and ends up around 1200 to 1400 µS/cm for all experiments. As these two values seem to adjust around this range, independently from the input, most of the reaction is assumed to be finished after 48 hours. Saturation indexes for calcite between −0.5 and −0.2 support this hypothesis. The concentration of Ca^{2+} and HCO_3^- show similar trends but also prove that a complete balance has not been achieved. All concentrations approach comparable magnitudes of about 180 to 190 mg/l Ca^{2+} (9.0–9.7 meq/l) and around 900 mg/l HCO_3^- (14.5–15.4 meq/l) and a Ca/HCO_3 ratio around 0.6–0.7. As these values evolve in all tests, and the dissolution reaction of calcite is becoming the dominant process to determine hydrochemical facies, but a definite correlation to the higher metal concentration in the sewage cannot be proven. Because of the known better solubility of calcite compared to dolomite calculated total amounts of dissolved calcite of between 350 and 400 mg/l are considerably higher in this experimental set.

Results of the long term pressure cell experiments show dolomite dissolution (and precipitation) effects as well, but because of the much lower reaction surface all absolute values are much smaller and as can be seen on Figures 2.6 and 2.7 this process is not straightforward.

In the first 650 hours experimental time of the test with natural water (Figure 2.6), Ca^{2+} values drop strongly and seem to show carbonate precipitation. As Mg^{2+} concentrations do not behave similarly the precipitated mineral must be calcite and the required bicarbonates come from the input solution. Their concentrations could not be measured due to low sample volumes at the beginning of the experiment. In a second phase from around 650 to 3000 hours, Ca^{2+} concentrations rise strongly, representing dolomite dissolution along the artificial fissure, accompanied by elevated pH around 8 and slightly rising EC and HCO_3^- values. As Mg^{2+} is always in a range of 2.5 to 3 meq/l, alternating to lower extents, this dissolution seems to be incongruent. After 3000 hours experimental time EC, Ca^{2+} and HCO_3^- concentrations show a distinct drop and indicate calcite precipitation.

In the experiment with the anthropogenic water type (Figure 2.7) Ca^{2+} and Mg^{2+} concentrations evolve together and seem to be affected by similar processes. Between

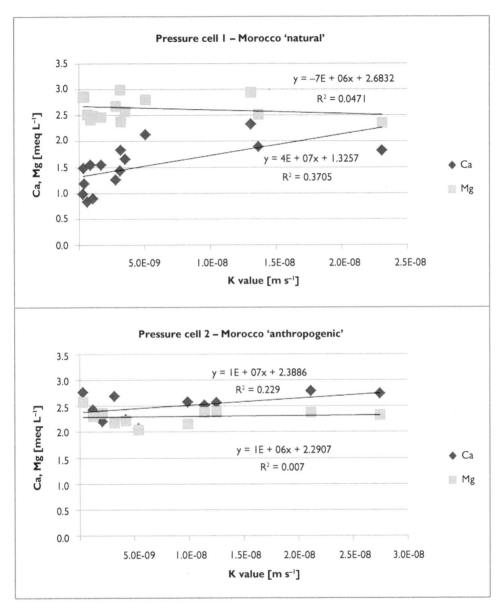

Figure 2.8 Correlation of measured Ca^{2+} and Mg^{2+} to hydraulic conductivity of the cut Moroccan rock samples.

around 400 and 650 hours concentrations of both cations decrease simultaneously, appearing to represent dolomite precipitation; HCO_3^- concentrations could not be measured to support this hypothesis. After that time values increase again accompanied by a rise in pH, typical for carbonate dissolution. At around 1500 hours all ionic concentrations show another obvious decrease, representing carbonate precipitation, but this time even with an excess in Mg^{2+} of around 0.2 meq/l. This periodic alternation of dolomite dissolution and precipitation seems to proceed to the end of the experimental time. As dolomite precipitation has not been reported for natural groundwater or experimental systems in situ this might be due to the added small amounts of heavy metals.

Because of the experimental setup in the pressure cells water flow and discharge through the fissure samples was not completely constant throughout the whole time. Small changes in applied pressure rapidly changed the aperture of the fissures and thereby also the flow velocities and K values respectively. As the artificial fissures in the rock samples are quasi-planar and all the water flows within them, the fissure apertures can be calculated from the the cubic law (e.g. Nicholl *et al.*, 1999). Values vary from 2.2 μm to 26.6 μm; below a critical value of 1 μm to the end of the second experiment hardly any water is able to pass the rock core. Corresponding K values during the experiments are between 2.9×10^{-10} and 5.1×10^{-7} m/s, and, therefore, up to five times higher than in the matrix. Although it is known that flow velocity effects carbonate dissolution kinetics (e.g. Morse & Arvidson, 2002), this seems not to be applicable to the present study. There is no obvious correlation to the measured Ca^{2+} and Mg^{2+} concentrations (Figure 2.8), so changes in their concentrations are dominated by hydrochemical properties.

2.5　CONCLUSIONS

Some conclusions can be drawn regarding the dissolution of calcite and dolomite in water under the influence of anthropogenically induced metal cations in short and long term experiments.

In batch tests with large reaction surfaces the dissolution of both carbonate minerals is a fast process leading to a strong release of Ca^{2+} and Mg^{2+} to the solvent. Dolomite dissolution is considerably slower than that of calcite and mainly an incongruent process providing an excess in Ca^{2+} compared to Mg^{2+}. Small amounts of added metal cations do not seem to have much influence on this process, as the differences are probably more related to slightly lowered pH values.

Long term experiments with cut hard rock samples of Moroccan dolomite represent groundwater flow through open fissures. Here reaction rates are much slower and during the experimental time of six months there are phases of incongruent dolomite dissolution and calcite precipitation in the natural water type. Under the influences of added metal cations the dolomite dissolves congruently and even dolomite precipitation seems to happen inside the fissure.

REFERENCES

Amraoui F., Razack M. & Bouchaou L.H. (2005) Impact of a long drought period on a large carbonate aquifer: the Liassic aquifer of the Sais plain and Middle Atlas plateau (Morocco). *Regional Hydrological Impacts of Climatic Change-Hydroclimatic Variability, IAHS Publ.* 296, 184–193.

Appelo C.A.J. & Postma D. (2005) *Geochemistry, Groundwater and Pollution.* Leiden, New York, 649p.

Ford D. & Williams P. (2007) *Karst Hydrogeology and Geomorphology*. Chichester, 562p.

Hamidi E.M., Boulangé B. & Colin F. (1997) Altération d'un basalte triasique de la région d'Elhajeb, Moyen Atlas, Maroc. *Journal of African Earth Sciences* **24**, 141–151.

Henningsen D. & Katzung G. (2002) *Einführung in die Geologie Deutschlands*. Heidelberg/Berlin, 214p.

Herman J.S. & White W.B. (1985) Dissolution kinetics of dolomite: Effects of lithology and fluid flow velocity. *Geochimica et Cosmochimica Acta* **49**, 2017–2026.

Kohlhaas W. (1972) Geologie, Hydrogeologie und Wasserhaushalt des Massenkalkes im nördlichen Sauerland und Bergischen Land (Remscheid-Altenaer Sattel, Herzkamper Mulde, Velberter Sattel, Rheinisches Schiefergebirge). Dissertation, RWTH Aachen, 119p.

Liu Z., Dreybrodt W. (1997) Dissolution kinetics of calcium carbonate minerals in $H_2O–CO_2$ solutions in turbulent flow: The role of the diffusion boundary layer and the slow reaction $H_2O + CO_2 \leftrightarrow H^+ + HCO_3^-$. *Geochimica et Cosmochimica Acta* **61**, 2879–2889.

Morse J.W. & Arvidson R.S. (2002) The dissolution kinetics of major sedimentary carbonate minerals. *Earth-Science Reviews* **58**, 51–84.

Nicholl M.J., Rajaram H., Glass R.J. & Detwiler R. (1999) Saturated flow in a single fracture: Evaluation of the Reynolds equation in measured aperture fields. *Water Resources Research* **35**, 3361–3373.

Zachara J.M., Cowan C.E. & Resch C.T. (1991) Sorption of divalent metals on calcite. *Geochimica et Cosmochimica Acta* **55**, 1549–1562.

Chapter 3

Hydrogeochemical distribution of Ca and Mg in groundwater in Serbia

Petar Papić, Jovana Milosavljević, Marina Ćuk &
Rastko Petrović
Faculty of Mining and Geology, University of Belgrade, Belgrade, Serbia

ABSTRACT

Ca and Mg are chemical elements commonly found in the environment and the main constituents of many types of minerals and rocks. They are also essential to Man. Owing to their abundance in nature, they are present in all water resources and generally occur as the dominant cations with low TDS levels, whose origin is associated with large formations of sedimentary rocks (limestones, dolomites), and to a lesser extent with the degradation of silicate minerals that contain Ca and Mg. Ca and Mg concentrations in groundwater in Serbia vary over a wide range and their distribution is not uniform. The variation in the concentrations of these ions depends on the hydrogeological province, while in any single province it is a consequence of Serbia's highly complex geology. The best examples are the Carpatho-Balkanides, with predominant karstified rock formations, and the Vardar Zone where ophiolites prevail but the structure is much more complex than that of the Carpatho-Balkanides.

3.1 INTRODUCTION

Ca and Mg are lithophile elements that play an important role in the composition of groundwater and in the environment in general, and are also essential to the human body (Jovic & Jovanovic, 2004). Ca is the most abundant alkaline earth metal in the Earth's crust and the main ingredient of many minerals and rocks. The most common Ca-rich minerals are: calcite, aragonite, dolomite, gypsum, apatite, clinopyroxenes, plagioclases, hornblende and epidote (Clare & Rhodes, 1999). According to estimates, the lithosphere contains 16.2–19.3 mg-Ca/g. The highest concentrations have been recorded in carbonate rocks (limestones, dolomites), and basalts among igneous rocks (Hitchon *et al.*, 1999). As rocks weather, Ca dissolves readily and enters the hydrosphere. The carbonate equilibrium is the main driver that restricts Ca migration in natural waters. In solution, Ca occurs as a bivalent ion, Ca^{2+} (Hitchon *et al.*, 1999).

Mg is a significant component of most rock formations and an important ingredient in many petrogenic minerals, such as dark ferromagnesian minerals (olivine, pyroxenes, amphiboles), but also minerals like serpentinite, talc, brucite, chlorite, biotite, tourmaline, dolomite, magnesite and spinel (Jovic & Jovanovic, 2004). Estimates of Mg concentrations in the lithosphere vary from 132 to 158 mg/g; the highest Mg concentrations tend to be found in ultramafic rocks (Hitchon *et al.*, 1999). In a sedimentary environment, Mg largely occurs in association with the carbonate ion, predominantly as dolomite $CaMg(CO_3)_2$. As a result of rock weathering, Mg^{2+} usually enters the hydrosphere as dark ferromagnesian minerals (e.g. chlorite, Mg-calcite and dolomite) degrade. In unpolluted shallow groundwater, Mg concentrations range from 0.1–1.2 to about 50 mg/l (Cox, 1995).

Ca and Mg are essential elements to Man and their impact on human health is enormous. Ca is found in the human body more than any other mineral. It takes part in the formation of bones and teeth. Muscle activity and the transmission of nerve impulses rely on Ca. It is valuable in blood coagulation, cardiac activity and enzyme production. In conjunction with Mg, it supports the function of the human heart. Lack of Ca increases the risk of high blood pressure and heart failure, while a prolonged deficit may lead to osteoporosis. Mg plays a multiple role in the human body: it acts directly on the neuromuscular plate, is essential for normal vitamin C and vitamin B_1 activity, takes part in enzymatic processes leading to energy production, reduces coagulation levels, protects the inner walls of blood vessels from fibrosis, and catalyses the utilisation of fats, proteins and carbohydrates (Teofilovic *et al.*, 1999; Cotruvo *et al.*, 2009).

Serbia is a country located at the crossroads of Central and South East Europe, covering the southern part of the Pannonian Plain and the central Balkans, lying between a massif and the Carpathian Mountains in the east, the Dinaric Alps in the west, and the Morava Valley – an intersection of land routes which lead southwards, towards Salonika, and eastwards, towards Asia Minor. The country is landlocked and borders Hungary to the north; Romania and Bulgaria to the east; Macedonia to the south; and Croatia, Bosnia, and Montenegro to the west; it also borders Albania through the disputed region of Kosovo. The capital of Serbia, Belgrade, is among Europe's oldest cities and one of the largest in East Central Europe.

The geology of Serbia is highly complex and not conducive to generalised assessments. Broad geological provinces have been identified based on geotectonic units. In general terms, the provinces are (Filipovic *et al.*, 2005): the Carpatho-Balkanides, the Serbian Crystalline Core, the Vardar Zone, the Inner Dinarides and the Pannonian Basin (Figure 3.1).

3.2 APPROACH AND METHOD

The data used in this research were derived from investigations conducted between 2008 and 2012. Groundwater was sampled at 257 locations across Serbia, including groundwater resources featuring low and high Total Dissolved Solids (TDS). The sampling network was designed to cover the entire territory of Serbia evenly and address groundwater occurrences in different rocks (igneous, metamorphic and sedimentary), and consequently different types of aquifers. The sampling points included springs, boreholes and wells. Sampling was conducted in accordance with the Drinking Water Sampling and Laboratory Analysis Rulebook (Official Gazette of the SFRY, no. 33/87). All groundwater samples were tested to determine the main physicochemical parameters (temperature, pH, electrical conductivity) and the basic chemical composition. The analyses were conducted at the Hydrochemistry Lab of the University of Belgrade Faculty of Mining and Geology, as well as at the Public Health Institute of Belgrade. Ca and Mg concentrations were determined by the ICP-OES method.

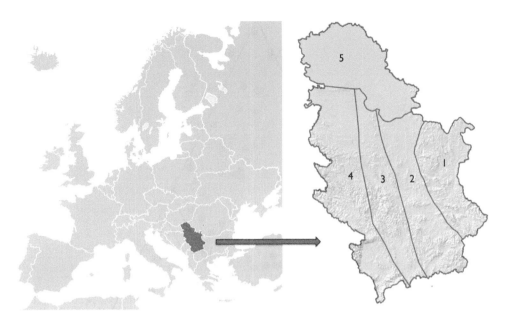

Figure 3.1 Geographic location and hydrogeological provinces of Serbia. Legend: 1. Carpatho-Balkanides; 2. Serbian Crystalline Core; 3. Vardar Zone; 4. Inner Dinarides; 5. Pannonian Basin.

Chemical analyses of groundwater samples were statistically processed to assess and interpret hydrochemical data and to generate hydrochemical maps of Ca and Mg distribution in the groundwaters of Serbia. The data were statistically processed and graphically interpreted using statistical software IBM SPSS v.19. The hydrochemical maps of the distribution of Ca and Mg, scale 1:500 000, were generated using ESRI ArcGIS 10.0 software.

3.3   RESULTS AND DISCUSSION

3.3.1   General groundwater quality

Serbia's highly complex geology has resulted in groundwater resources featuring different types, temperatures and TDS levels. The dominant anion was the hydrocarbonate ion. Apart from several occurrences of sulfate, chloride, hydrocarbonate-sulfate and hydrocarbonate-chloride types, more than 90% were found to be of the hydrocarbonate type of groundwater. Of all the samples, three belonged to the sulfate type and two to the chloride group with a chloride portion of 97% equivalent. The latter two occurrences show high TDS levels, in excess of 6000 mg/l. Based on their cation composition, the samples predominantly reflected Ca, Na and composite (Ca–Na, Ca–Mg, Ca–Mg–Na) types of groundwater. Four samples were of the Mg type, with a Mg component in excess of 75% equivalent (Figure 3.2).

With regard to Total Dissolved Solids (TDS), the samples exhibited considerable diversity: from low levels (only 29 mg/l) to very high levels (in excess of 20 000 mg/l).

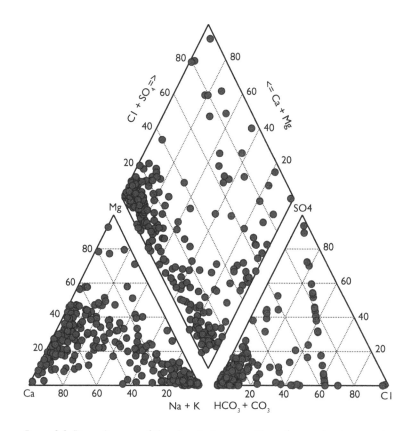

Figure 3.2 Piper diagram of the chemical composition of groundwater samples.

3.3.2   Calcium in the groundwaters of Serbia

Ca concentrations in Serbia's groundwaters were found to vary over a broad range, depending on the type of groundwater and TDS: from 0.60 mg/l to 392.80 mg/l. The median was 82.32 mg/l.

Ninety-five or 37% of the samples had Ca concentrations up to 50 mg/l, while the majority of the samples, 107 or 42%, contained Ca in concentrations ranging from 51 to 100 mg/l, meaning that some 80% of samples featured Ca in concentrations up to 100 mg/l (Figure 3.3).

Of the 257 occurrences groundwater samples, only 20 (8%) featured Ca concentrations in excess of the Maximum Allowable Concentration (MAC), which is 200 mg/l according to national drinking water standards (Figure 3.3). In view of the TDS levels, high Ca concentrations are common, especially in carbonated groundwater, given that elevated CO_2 concentrations enhance the solubility of Ca carbonates and, consequently, increase Ca concentrations in the groundwater.

As Ca is abundant in the Earth's crust, it is found in all natural water resources. Even though Ca^{2+} mostly relates to sedimentary rocks (limestone, dolomite), it is difficult to explain that high Ca^{2+} concentrations have been recorded in groundwater

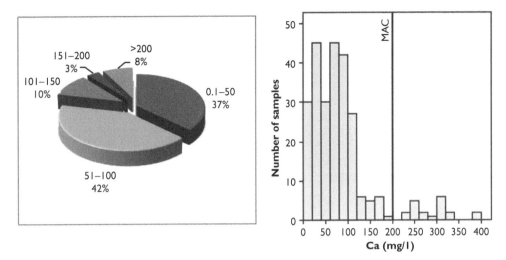

Figure 3.3 Plot of relative frequency of Ca and histogram of Ca concentrations in Serbia's groundwaters.

occurring in limestones because Ca concentrations in groundwater are limited by the solubility of calcium-carbonates. High Ca concentrations in water, typical of mineral water, are conditional upon the presence of other components (CO_2 and other ions), which affect the concentrations of dissolved mineral substances that together contribute to the TDS level of the water.

Ca is not evenly distributed in Serbia's groundwater resources but certain relationships apply with the respective geological makeup and are apparent in the various hydrogeological provinces (Figure 3.4).

For example, the most uniform distribution of Ca in groundwater is found in the Carpatho-Balkanides, which is as expected given that groundwater in this province occurs in predominant karstified rock formations. Ca concentrations are typical of the low-TDS HCO_3–Ca type of groundwater, with an average of about 75 mg/l. One sample had a Ca concentration of 240 mg/l; this was carbonated groundwater with a TDS level of 1294 mg/l, whose origin is rather complex and governed by tectonics and the geological makeup of the rocks underlying the karstified formations. The presence of a hidden granite intrusion has caused the generation of CO_2, enhanced the solubility of calcium-carbonates and enriched this groundwater with Ca.

The Inner Dinarides of western Serbia also feature certain regularities in the distribution of calcium. Groundwater occurrences originally associated with karstified formations were largely found to be of the low-TDS HCO_3–Ca type, with calcium concentrations up to 70 mg/l. Contrary to this type, high-TDS groundwater occurrences whose calcium concentrations measured up to 380 mg/l were associated with flysch sediments and schists. The origin of such groundwater is intricate and tectonics plays an important role in the formation of its chemical composition.

Within the Pannonian Basin, groundwater is of the hydrocarbonate type, with hydrocarbonate ion concentrations in excess of 50% equivalent. Their cation

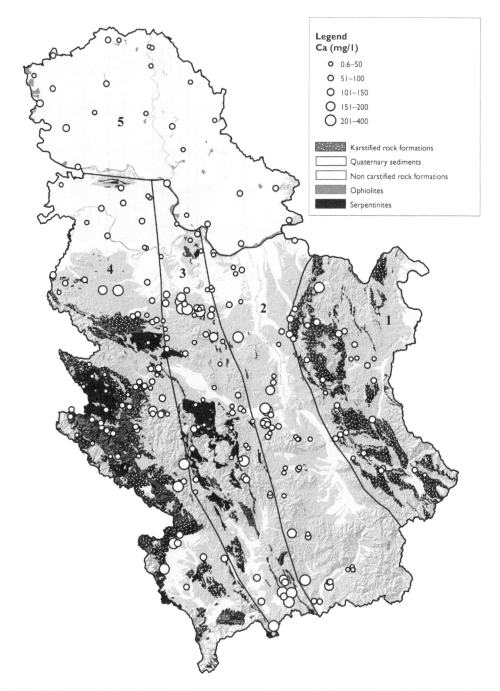

Figure 3.4 Ca distribution in Serbia's groundwaters. Legend: 1. Carpatho-Balkanides; 2. Serbian Crystalline Core; 3. Vardar Zone; 4. Inner Dinarides; 5. Pannonian Basin.

Table 3.1 Ca concentrations in different hydrogeological provinces (mg/l).

Province	Number of samples	Minimum	Average	Maximum
Carpatho-Balkanides	27	14.00	75.86	240.00
Serbian Crystalline Core	54	0.60	61.22	310.00
Vardar Zone	62	3.00	67.02	392.80
Inner Dinarides	85	1.80	66.66	380.00
Pannonian Basin	29	2.40	53.25	138.00

composition is complex, dominated by Ca, Mg and Na ions in different ratios. Table 3.1 shows that the concentration range in this province is the smallest. Higher concentrations were noted in certain high-TDS samples, such as the occurrence of the Cl–Na type of groundwater with a TDS level of 6600 mg/l, where the Ca concentration was 114.9 mg/l but it constituted only 0.7% of the cation composition.

The largest variations in Ca concentrations were noted in the Serbian Crystalline Core and the Vardar Zone, as a result of the complex geology of Central Serbia. In these provinces concentrations ranged from only 0.6 mg/l to as much as 392.8 mg/l (the highest recorded concentration). This groundwater is of the HCO_3–Mg–Ca–Na type, with a TDS level of nearly 5 g/l.

3.3.3 Magnesium in the groundwaters of Serbia

Mg concentrations in Serbia's groundwaters vary considerably, from 0.05 mg/l to 378 mg/l. The average is 34.27 mg/l. Mg concentrations depend on the geology and tectonics, as well as the type of groundwater and the TDS level, given that the concentration of this ion is higher in high-TDS than in low-TDS groundwater, and that this groundwater is not of a pure Mg type.

In most of the samples (106 or 41%), Mg concentrations were only up to 15 mg/l; 26% or 66 samples contained 15 to 30 mg/l, while 15% (39 samples) were between 30 to 50 mg/l. This means that 82% of the groundwater samples collected across Serbia had Mg concentrations up to 50 mg/l, which is the maximum allowable concentration according to national drinking water standards.

Of the 257 samples, 45 samples (or 18%) had Mg concentrations in excess of drinking water standards (Figure 3.5). Compared with Ca concentrations (only 20 of 257 samples exceeding the MAC), Mg had a larger number of exceedances but in view of the total number of samples analysed and the fact that the MAC for Mg is considerably lower than that for Ca, such a proportion was expected.

Mg concentrations in excess of 50 mg/l were found in high-TDS groundwater of up to 6000 mg/l. Such groundwater is generally traced to schists. In several samples of low-TDS groundwater, Mg concentrations measured 50–60 mg/l; such groundwater is largely originally associated with dolomite and dolomitic limestone, as well as interfaces of these rocks with flysch, Neogene sediments, fractured and degraded sandstones and marls, or rocks whose composition includes Mg-rich minerals.

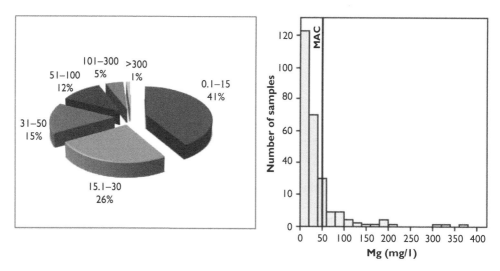

Figure 3.5 Plot of relative frequency of Mg and histogram of Mg concentrations in Serbia's groundwaters.

Occurrences of low-TDS groundwater, where Mg concentrations are in the 60–70 mg/l range, were associated with serpentinites, fractured harzburgites and dolomite/serpentinite interfaces.

As with Ca, the distribution of Mg is not uniform (Figure 3.6), but regularities resulting from the geological makeup are much more apparent.

Table 3.2 shows the smallest variations but also the lowest Mg concentrations in the Carpatho-Balkanide Province, where the average Mg concentration was found to be 16.52 mg/l. One sample from this province exceeded the MAC (52 mg/l), but this was a result of the influence of flysch in the vicinity, even though the province is dominated by limestone and dolomitic karstified rock formations, as corroborated by the Mg to Ca ratio.

The Inner Dinarides of western Serbia contained an average Mg concentration in groundwater of 25.29 mg/l; the concentrations of this ion were generally below the MAC, although there were several exceptions where concentrations were considerably higher (up to 378 mg/l). The occurrences that gave Mg concentrations up to the average level (about 25 mg/l) were generally low-TDS groundwater resources from the limestone formations of this province, with a slight influence of diabase-chert rocks or Neogene sediments. Mg concentrations from the average level to 50 mg/l were recorded in samples collected from groundwater occurrences in limestones influenced by serpentinites, harzburgites and similar rocks that make up the Dinaride ophiolite zone. This province featured a distinct occurrence of groundwater of the Mg type associated with pure serpentinites and harzburgites. This was low-TDS groundwater (350 mg/l), in which the Ca concentration was very low Ca (8.02 mg/l) and the Mg concentration was not high (66.63 mg/l), though it comprised 92.7% of the cation composition. The highest concentration of 378 mg/l was recorded in a sample that

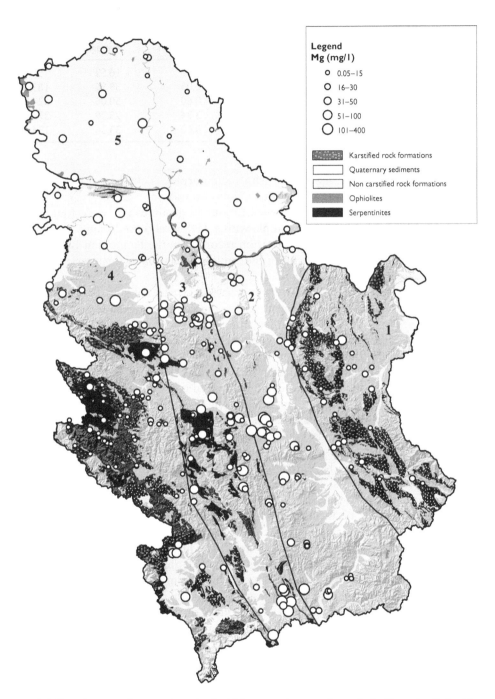

Figure 3.6 Distribution of magnesium in Serbia's groundwaters. Legend: 1. Carpatho-Balkanides; 2. Serbian Crystalline Core; 3. Vardar Zone; 4. Inner Dinarides; 5. Pannonian Basin.

Table 3.2 Mg concentrations in different hydrogeological provinces (mg/l).

Province	Number of samples	Minimum	Average	Maximum
Carpatho-Balkanides	27	1.00	16.52	52.00
Serbian Crystalline Core	54	0.05	39.95	183.00
Vardar Zone	62	0.60	50.06	324.00
Inner Dinarides	85	3.24	25.29	378.00
Pannonian Basin	29	5.60	33.36	194.00

had a TDS of 4995 mg/l, associated with metamorphic rocks (slates, phyllites) and sandstones, which also had the highest Ca concentration.

In the Pannonian Basin, Mg concentrations in groundwater were found in the range 5.6 mg/l to 50.5 mg/l. One sample with a TDS level of 6620 mg/l had 194 mg-Mg/l, which was the highest concentration recorded in the Pannonian Basin. Such a high Mg concentration was a result of the aquifer formation at the interface between serpentinites and Tertiary sediments.

Unlike the other provinces, where only a few occurrences of high-TDS groundwater were found to contain high Mg concentrations, the provinces of the Serbian Crystalline Core and Vardar Zone have the largest number of occurrences with Mg concentrations exceeding 50 mg/l (MAC for drinking water). In the Serbian Crystalline Core, Mg concentrations were found to range from 0.05 mg/l (the lowest recorded in Serbia) to 183 mg/l, with an average concentration of 39.95 mg/l. Fourteen samples had Mg concentrations above 50 mg/l and they could be divided into two groups: (1) groundwater occurrences with Mg concentrations are between 53 and 70 mg/l and whose composition is influenced by schists, metasandstones and Neogene sediments containing Mg-rich silicate minerals, and (2) occurrences of high-TDS groundwater (1523 to 4995 mg/l) associated with schists, whose Mg concentrations measured from 80 to 183 mg/l.

Mg concentrations in the Vardar Zone groundwater were found to be higher than in the other four provinces. They ranged from 0.6 mg/l to 324 mg/l. The average was 50.06 mg/l, which was the highest average Mg concentration among the provinces. Two groups of groundwater occurrences in the Vardar Zone can be identified: (1) groundwater occurrences with Mg concentrations up to MAC (50 mg/l), or up to the average Mg concentration for this province, generally related to limestones and influence by flysch formations, schists and Neogene sediments, and (2) 20 groundwater occurrences whose Mg concentrations were in excess of 50 mg/l, where as many as seven samples had Mg concentrations above 100 mg/l, and two samples were in excess of 300 mg/l. Such high Mg concentrations are likely a result of the extent of the ophiolitic belt of the Vardar Zone, as suggested by the rMg/rCa ratio. Vardar Zone ophiolites also influence Mg concentrations of the first group of groundwater occurrences in this province, formed not in ophiolites but other rock formations (primarily karstified Triassic limestones interchanging with less pervious diabase-chert formations and ultramafic rocks).

3.3.4   Mg/Ca ratio in groundwater of Serbia

The Mg/Ca ratio of groundwater is important because it is an indicator of the lithological composition of the aquifer matrix. For instance, a Mg/Ca ratio of 0.7 may

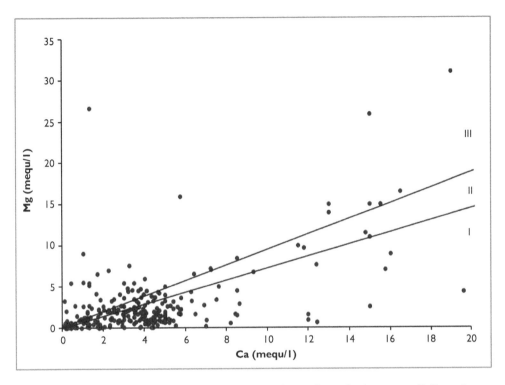

Figure 3.7 Mg/Ca ratio of groundwater. Legend: I. Groundwater formed in limestones; II. Groundwater associated with dolomites and dolomitic limestones; III. Groundwater tracing to Mg-rich silicate rocks (ophiolites and ultramafic rocks).

suggest that groundwater was formed in limestones. Groundwater occurrences with an Mg/Ca ratio of 0.7–0.9 are generally associated with dolomitic limestones, while an Mg/Ca ratio greater than 0.9 is indicative of groundwater from Mg-rich silicate rocks. If the ratio is greater than 1, the groundwater relates to ophiolites and ultramafic rocks, as well as ophiolitic detritus in the sediments (Mandel & Shiftan, 1981).

The Mg/Ca ratio of the groundwater samples in Serbia was found to range from 0.02 to 36.62, suggesting diverse lithological compositions and complex geology, or groundwater occurrences relating to a variety of rock types. Based on the Mg/Ca ratio, 58% of the groundwater occurrences related to limestones, only 9% to dolomites, and 33% to silicate rocks, i.e. 25% to ophiolites and ultramafic rocks (Figure 3.7).

The importance and accuracy of classification of the types of rocks in which groundwater is formed, based on the Mg/Ca ratio, is best demonstrated by the Carpatho-Balkanide Province. There, 30% of the province feature karstified rock formations, and karst aquifers. According to the Mg/Ca ratio, 24 out of the 27 groundwater samples from this province related to limestones, as corroborated by the geological makeup. Two samples had Mg/Ca ratios of 0.7 and 0.9, and their chemical composition was under the influence of dolomites in conjunction with limestones. Only one sample had a ratio of 0.91, suggesting the presence of silicate rocks and corroborated on the ground by andesites and flysch sediments in part of this province, along with dominant Cretaceous limestones. This groundwater is influenced by silicate rocks but the aquifer was not contained in them, such that the ratio is close to the limit of 0.9 for this group. Otherwise it would be much higher.

Looking at the calculated Mg/Ca ratios compared with the geological makeup, it is apparent that the Mg/Ca ratio of the rock type in which the groundwater occurs matches the geological makeup, and that this ratio can be used to determine the origin of groundwater or at least to narrow down the list of possibilities if data on the geological environment are not available; this parameter can be used to determine the effect of lithology on the formation of the chemical composition of groundwater.

3.4 CONCLUSION

It is evident that Ca and Mg concentrations in Serbia's groundwaters vary over a wide range: 0.6 to 392.8 mg/l for Ca and 0.05 to 378 mg/l for Mg. Uneven distributions and large differences in concentrations have been noted not only between provinces, but also within a single province, as a result of complex geology, attesting to the fact that lithology is the main driver of the chemical composition of the groundwater. Additionally, Total Dissolved Solids (TDS) are a significant parameter as the concentration of one or both of these ions in high-TDS groundwater is considerably higher than in low-TDS groundwater.

Analyses of the Mg/Ca ratio of groundwater in Serbia and the identification of the types of rocks in which groundwater occurs, revealed, based on specified theoretical values, that these ratios largely matched the geological makeup. It was, therefore, safe to conclude that the Mg/Ca ratio may be used as a parameter for tentative, not definitive, identification of the types of rocks that had a dominant influence on the formation of the chemical composition of the groundwater.

The Mg/Ca ratio is also an important drinking water parameter and the recommended (ideal) ratio of these two ions in water is 1:2. Given that, according to the recommended Mg/Ca ratio more than 60% of the groundwaters in Serbia originate from limestones and that quite a few of them exhibit the ideal ratio, but such groundwater is precious from a drinking water perspective.

ACKNOWLEDGEMENT

This research was supported by the Ministry of Education, Science and Technological Development, Government of Republic of Serbia, through Project III 43004.

REFERENCES

Clare M. & Rhodes F. (1999) *Encyclopedia of Geochemistry*. Kluwer Academic Publishers, Dordrecht, The Netherlands.

Cotruvo J. & Bartram J., (editors) (2009) *Calcium and Magnesium in Drinking water: Public health significance*, World Health Organization, Geneva.

Cox P.A. (1995) *The Elements on Earth – Inorganic Chemistry in the Environment*. Oxford University Press, Oxford.

Filipovic B., Krunic O. & Lazic M. (2005) *Regional hydrogeology of Serbia*, Faculty of Mining and Geology, Belgrade.

Hitchon B., Perkins E.H. & Gunter W.D. (1999) *Introduction to Ground Water Geochemistry*. Geosciene Publishing Ltd, Sherwood Park, Alberta.

Jovic V. & Jovanovic L. (2004) *Geochemical basis of ecological management*. Society for the Dissemination and Application of Science and Practice in Environment of Serbia and Montenegro "Ekologica", Belgrade.

Mandel S. & Shiftan Z. (1981) *Groundwater Resources, Investigation and Development*. Academic Press New York & London.

Regulation of Hygienic correctness of Drinking water, Official Gazette of SRY, no. 42/98 and 44/99.

Teofilovic M., Obrenovic A. & Pesic D. (1999) *Importance of calcium (Ca) and magnesium (Mg) in nature – spatial distribution and their content in mineral waters of Serbia*. EKOLOGICA 6, no. 4, Belgrade.

Chapter 4

Study of Ca and Mg distribution within the aeration and saturation zones of the Upper Jurassic limestone massif of the southern part of the Cracow-Częstochowa Upland (Poland)

J. Różkowski[1] & K. Różkowski[2]

[1]Faculty of Earth Sciences, University of Silesia, Sosnowiec, Poland
[2]Faculty of Mining and Geoengineering, AGH University of Science and Technology, Kraków, Poland

ABSTRACT

The Cracow-Częstochowa Upland is the largest karst area in Poland. Within its borders is a fissure-karstic aquifer associated with the Upper Jurassic carbonate rocks. This paper presents directions of infiltration water chemical evolution, especially variability of Ca and Mg concentrations in precipitation (rain and snow), cave waters of the vadose zone as well as waters of the phreatic zone (springs and wells) sampled at the Cracow Upland. The processes of carbonate leaching based on hydrogeochemical modeling are described, taking into account the chemical and mineral composition of the Upper Jurassic limestones. The study was carried out at the research areas of Prądnik catchment between 1995 and 2006 and on the Zakrzówek horst between 1996 and 1997.

4.1 INTRODUCTION

This chapter presents the distribution of Ca and Mg in groundwater in the Cracow-Częstochowa Upland, the most extensive karstic area of Poland. Within this Upland is a fissure-karstic-porous aquifer of Upper Jurassic age. Great thickness of an aeration zone and small thickness of shallow phreatic zone result from deep cut of karstic aquifer. Investigations were conducted between the years 1995 and 2006 within the aeration and saturation zone of the Prądnik river catchment, 30 km north west of Cracow (in the Ojców National Park) and between 1996 and 1997 in the aeration zone of the Zakrzówek horst impacted by anthropogenic processes from the city of Cracow (Figure 4.1) (Różkowski, 2006, 2008).

The Upper Jurassic aquifer of the Cracow Upland is located in a carbonate massif which is heterogeneous, discontinuous and anisotropic, and is a triple porosity medium. Karst fissures and channels (subjected to hierarchical processes) are the main flow paths while pore spaces favour water storage. Recharge of the aquifer, mainly of dispersed character, occurs over the entire outcrop area and it also takes place

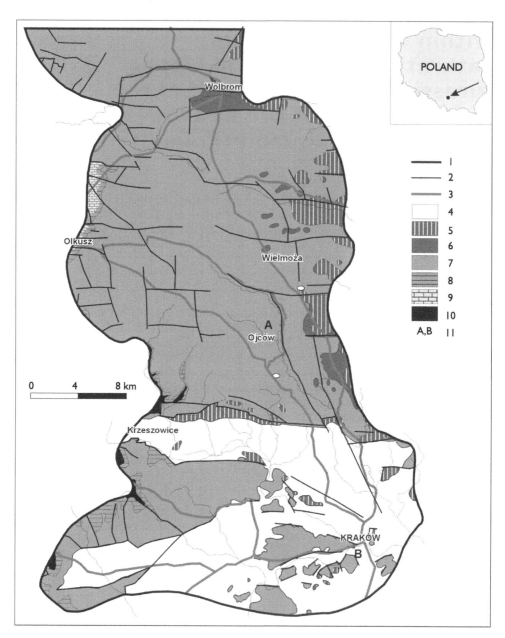

Figure 4.1 Geological map of the Cracow Jurassic (without Quaternary sediments, acc. to Jóźwiak, Kowalczewska, 1984, modified) 1 – limits of the Upper Jurassic aquifer (acc. to Paczyński ed., 1995); 2 – main faults; 3 – roads; sediments: 4 – Neogene: clays, claystones, silt-stones, sands, sandstones, gypsum, 5 – Upper Cretaceous: marls, limestones, 6 – Lower Cretaceous: sands, sandstones, conglomerates, 7 – Upper Jurassic: limestones, marls, 8 – Middle Jurassic: clays, mudstones, conglomerates, marls, limestones, 9 – Triassic: clay-stones, dolomites, 10 – Carboniferous and Devonian: claystones, siltstones, sandstones, hard coal, limestones, marls, dolomites, 11 – study area: A – Prądnik river catchment, B – Zakrzówek horst.

indirectly through the overlying Quaternary deposits. Local and intermediate flow systems dominate with discharge directed to deeply incised river and stream valleys. The unsaturated zone is thick. The flow system in the Cracow Jurassic strata reacts mostly with a substatial delay to precipitation events and is classified as 'Torcal type' (Różkowski *et al.*, 1997).

4.2   PRESENCE OF Ca AND Mg IN SOIL
IN THE RESEARCH AREA

The concentration of Ca in the soil in the research area ranges from 0.5 to 1.0 wt%, while in soil developed over the Jurassic limestone it is less than 1 wt% and at the regions of high human impact reaches 2 to 4 wt%. Baseline values of Ca in soils in Poland are up to 0.5 wt%. The spatial distribution of Mg in soils within the research area is similar to the distribution of Ca. Determined values range from 0.05 to 0.1 wt%, while in the areas influenced by human activity the range is between 0.1 and 0.2 wt%. Baseline values of Mg in soils is lower, reaching 0.2 wt%. Observed enrichment of Ca and Mg in soils in the urbanised areas is associated mainly with precipitation of particulate matter generated by industry, originating mainly from coal burning (Lis & Pasieczna, 1995).

4.3   CHEMICAL AND MINERALOGICAL COMPOSITION
OF THE UPPER JURASSIC LIMESTONE

During the investigation of the chemical and mineralogical composition of the limestones, samples from 17 quarries and other outcrops along the Cracow-Częstochowa Upland were collected. Mineralogical studies were performed in the X-ray diffraction Laboratory of the Department of Earth Sciences, University of Silesia. On the basis of the geochemical studies, using X-ray diffractometer and atomic absorption techniques, it was found that the main mineral component – calcite is present in amount of 90–99% by volume, while magnesium varied between 1617 to 2439 ppm, corresponding to 0.16–0.24%. The average content of $CaCO_3$ and $MgCO_3$ in mineable limestone deposits was respectively 88 to 97% and 0.8 to 1.6%.

Based on the analysis of the chemical composition of samples selected from 12 Upper Jurassic limestone deposits, documented within the Cracow Jurassic, it was found that the average content of the basic components of the limestones are: $CaCO_3$ 88–97%, $MgCO_3$ 0.7–1.6%, SiO_2 0.8–8.6%, Fe_2O_3 0.1–0.8%, Al_2O_3 0.4–2.4%, SO_3 0.05–0.3%, Na_2O 0.03–0.1%, K_2O 0.05–0.3% (Różkowski, 2006).

Low magnesium calcite is the main component, while the secondaries are: quartz, clayey minerals, occasionally feldspars of potassium (sanidyn) and sodium (albit), volcanic glass, goethyt – hematite, pyrite (Bzowska *et al.*, 2008; Różkowski *et al.*, 2009). Close to the upper part of the Jurassic sediments the processes of silification, piritization and dolomitization were observed (Dżułyński & Żabiński, 1954).

4.4   CHEMISTRY OF GROUNDWATER IN THE UPPER JURASSIC CARBONATE ROCK MASSIF OF THE CRACOW UPLAND

Chemical analyses of 1100 samples of rainwater, infiltrating water in the vadose zone and groundwater in the phreatic zone were carried out in the laboratories of the University of Silesia, and the AGH University of Science – Technology in Cracow, using among others atomic emission spectroscope with inductively coupled plasma ICP-AES Plasma 40 and mass spectroscope with inductively coupled plasma ICP-MS Elan. To determine the loads of the individual components in the hydrosphere, the results of numerical modeling studies carried for determine balance evaluation and groundwater renewal estimation in Prądnik catchment were used (Różkowski *et al.*, 2005).

The chemical composition of groundwater within the Upland area evolves, both in the vertical profile as well as during horizontal flow – from the recharge to the drainage zone. Quantitative analysis of the hydrochemical transformation in the vertical profile of the partial catchment of Prądnik river revealed a radical modification of infiltrating water in relation to the rainfall as a result of intensive leaching of the soil matrix and ion exchange processes. The load of Ca increases many times, while the load of Mg decreases in concentration in connection with vegetative and soil processes. The process of limestone dissolution during vertical percolation has a moderate dynamic which is reflected in a slight increase in the Ca and Mg concentrations within the phreatic zone in contrast to the vadose zone.

The average concentrations of Ca and Mg in precipitation within Prądnik river catchment are respectively 2.0 and 0.6 mg/l, in cave water in the vadose zone 98.0 and 1.2 mg/l, while they are 101 and 5.6 mg/dm^3 in samples from the phreatic zone. Assuming average precipitation of 0.00198 m/d, the volume of infiltration 518 m^3/d/km^2, and the total subsurface runoff 524 m^3/d · km^2 (deried from mathematical modelling), the loads of Ca and Mg in the hydrosphere in the Prądnik catchment are respectively 4.0 and 1.2 kg/d · km^2, 50.8 and 0.6 kg/d · km^2, 53.0 and 2.9 kg/d · km^2. Greater dynamics of the dissolution process than between the vadose and phreatic zones can be observed in the horizontal flow lines at the regional scale (total hardness increases from 200 to 325 mg CaCO$_3$/l).

Dissolving of the Upper Jurassic carbonate rocks in the saturation zone during the horizontal flow causes an increase in concentration of those components that are readily leached. The regional flow direction is from east towards the north east and from tectonic horsts towards tectonic troughs. The regular distribution of Ca, Mg, HCO$_3$, SO$_4$, Sr, SiO$_2$ in the groundwater, contrasts with the irregular distribution of anthropogenic origin elements Na, K, NO$_3$, PO$_4$. The modal values of these constituent concentrations in waters sampled from caves in the vadose zone within the karstic catchment of Prądnik river amount to Ca 90.0 and Mg 0.2 mg/l. The cave waters are dominated by HCO$_3$–SO$_4$–Ca, some also HCO$_3$–Ca chemical type, with conductivity ranging from 300 to 1100 μS/cm.

Modal values of the same constituents in waters in phreatic zone from the whole southern part of the Upland differs slightly, reaching the values Ca 96.0 and Mg 2.4 mg/l. Within the phreatic zone the specific electrical conductivity ranges from 300 to 1900 μS/cm and the water is of the HCO$_3$–Ca type, modified to HCO$_3$–SO$_4$–Ca and HCO$_3$–Cl–Ca–Mg in waters impacted by anthropogenic processes. The range of Ca and Mg concentrations in waters in the phreatic zone are respectively 24 to 243 and 0.2 to 53.3 mg/l. Ranges correspond to concentrations typical for natural groundwater, respectively from 2 to 200 and from 0.5 to 50 mg/l (Witczak &

Adamczyk, 1995). The coefficient of variation of Ca concentrations does not exceed 25% while for Mg it reaches 105%.

The chemical composition of the groundwater along flow paths within the local circulation systems indicates decreasing rCa/rMg coefficient variability from 76 to 5. At the same time the value of the coefficient points to the lack of dolomite, in the presence of which the indicator rCa/rMg attains values between 1.5 to 3.0.

The chemical evolution in the aeration zone in the Upper Jurassic limestones was investigated in detail in the Zakrzówek horst within the karst system impacted by anthropogenic processes from Cracow city (Figure 4.2). Hydrochemical

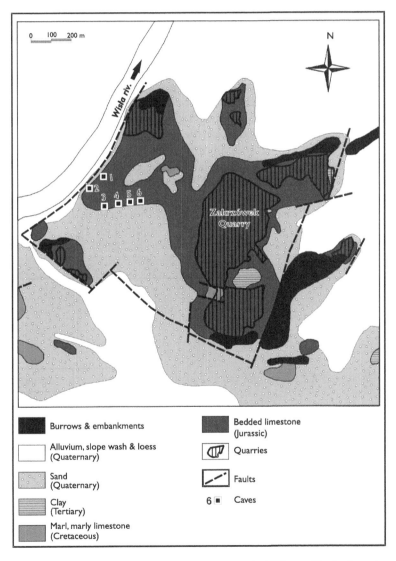

Figure 4.2 Sketch of Zakrzówek horst geology (Kulma *et al.*, 1991, modified)   Investigated Caves: 1 – Jasna, 2 – Wywiew, 3 – Z Kulkami, 4 – Niska, 5 – Pod Nyżą, 6 – Twardowskiego.

investigations comprised of 70 analyses of precipitation and 421 chemical analyses of percolating water from seven caves. Mean pH of precipitation was: rain 5.97, sleet 5.48, snow 4.93. Concentrations of Ca and Mg in precipitation were in the range Ca 0.4 to 26.7 mg/l (modal value 3.4) and Mg 0.05 to 4.1 mg/l (modal value 0.62) (Figure 4.3). The highest concentration of cations was in sleet (Ca 10.6 mg/l, Mg 1.14 mg/l), lower in rain (Ca 5.02 mg/l, Mg 1.03 mg/l) and lower still in snow (Ca 2.75 mg/l, Mg 0.39 mg/l). Among the hydrochemical types of precipitation are HCO_3–SO_4–Cl–Ca, SO_4–HCO_3–Cl–Ca, SO_4–HCO_3–Ca, NO_3–SO_4–Ca types while specific electrical conductivity increases from 7 to 405 μS/cm (modal value 38 μS/cm).

The presence of the Ca ion in precipitation is connected with porous particles of natural and anthropogenic dust. Particles of $CaCO_3$ from rock decomposition and weathering, especially in a carbonate environment, may be carried into air as dust. Calcium carbonate is also an important component of fly-ash. The presence of Mg in precipitation may be connected with aeolian transport of soil waste, fertilizer and wood burning (Johnson & Lindberg, 1989). Alluvial mineral particles mixed with snow cause a significant Ca presence in melting snow (Jeffries, 1990).

The reaction of water from the Zakrzówek horst aeration zone changed it from slightly acid (pH 6.82) in water remaining a short time in the massif to slightly alkaline (maximal pH 8.62). The acidity of the infiltrating precipitation was radically decreased as a result of calcite and subordinately dolomite dissolution in the presence of CO_2. Conductivity differed spatially in the caves from 130 to 2160 μS/cm. Ca predominated and Mg was secondary. The concentrations of Ca varied between 25 and 490 mg/l (modal value 76 mg/l) and Mg 0.07 and 90 mg/l (modal value 4.6 mg/l) (Figure 4.3). In caves in the Zakrzówek horst the hydrochemical types of water are HCO_3–SO_4–Ca, SO_4–HCO_3–Ca (64% of population).

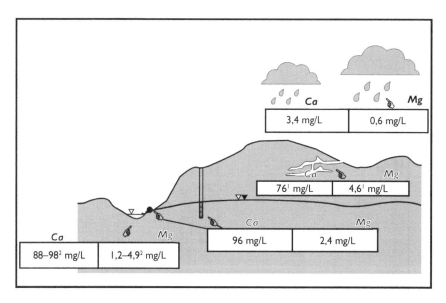

Figure 4.3 Ca and Mg ions in hydrosphere of Cracow Upland 1 – in Zakrzówek horst; 2 – mean concentrations.

4.5 PROCESS OF CARBONATE ROCK LEACHING IN HYDROGEOCHEMICAL NUMERICAL MODELING

In the Cracow Upland modification of fissure-karstic water chemistry is caused by a process of natural leaching of carbonate rock as seen in results of hydrogeochemical numerical modeling with the PHREEQC program (Parkhurst & Appelo 1999; Drever 1997). Saturation indexes (SI) for carbonate minerals in the Zakrzówek horst area calculated as a result of speciation modeling are saturated for calcite (SI_C from −0.32 to +0.13) and unsaturated for dolomite (SI_D from −3.01 to −0.83). Dissolution of dolomitised parts within the carbonate rock massif takes place (SI_D in precipitation is −9.73) but it does not significantly modify the chemical composition of water infiltrating into the massif. Ca^{2+} ion predominates among calcium species in the aeration zone water, and ionic pairs $CaSO_4^o$, $CaHCO_3^+$, $CaHPO_4^o$, $Ca\,H_2PO_4^+$ and $CaCO_3^o$ are present in traces.

The analyses, including the chemical composition of the phreatic zone groundwater, were undertaken along the regional and intermediate groundwater flow paths. Calculated SI as a result of hydrogeochemical speciation modeling vary for carbonate minerals around the values close to equilibrium state with groundwater, what is typical for diffuse flow system. This is a result of dissolution of carbonate rocks during the infiltration of acidic precipitation through the vadose zone. For all the flow paths, a gradual water saturation against predominating mineral phases in rocks is observed. The most prevalent forms of dissolved compounds in the recharge zones are simple ion forms (Ca^{2+} and Mg^{2+}). For all investigated directions of groundwater flow the trends were a decrease in the percentage of simple ion forms and an increase in the percentage of sulphates ($CaSO_4^o$, $MgSO_4^o$), associated with the dissolution of sulphates and the oxidation of sulphides and the dominance of acid carbonates $(Ca, Mg)HCO_3^+$ over the carbonate forms. The amounts of simple ion forms (Ca, Mg) and acid carbonates as well as sulphates show small seasonal variations. In the described environment of carbonates, exclusively carbonate forms reach the saturation state, while sulfate and gaseous forms (eg. CO_2) have a tendency to dissolve in water over the whole year.

4.6 CONCLUSIONS

In the hydrosphere of the Cracow Upland a hydrochemical transformation both in the vertical profile as well as during the horizontal flow of groundwater takes place from the recharge zone to the drainage zone. There is an increase in Ca and Mg concentrations in infiltrating waters, modified by human activity (Figure 4.3). This was observed during studies conducted in the Prądnik catchment located partially within the protected area of the Ojców National Park, as well as at the Zakrzówek horst situated in the Cracow city limits. In the urban area of Cracow the concentrations of Ca and Mg in precipitation between the years 1996–1997 ranged between Ca 0.4 and 26.7 mg/l and Mg 0.05 and 4.1 mg/l (modal values 3.4 mg/l and 0.6 mg/l). The highest cation concentrations were observed in sleet, were lower in a rain and lowest in a snow. Intensive dissolution of the soil matrix and the Upper Jurassic carbonates, in which low-Mg calcite content reaches 90–99% calcite, by acid precipitation (mean pH of rain 5.97, sleet 5.48, snow 4.93), within the waters of vadose zone causes a dramatic increase in the concentrations of Ca and Mg. Detailed studies conducted at the Zakrzówek horst show that the concentrations of cations in the cave waters vary significantly (Ca 25 to 490 mg/l and Mg 0.07 to 90 mg/l). This results from the complex circulation conditions in the fracture-karst porous carbonate massif and the locally strong anthropogenic impact.

Below the epikarst zone and the upper part of the vadose zone, the process of limestone dissolution by percolating water has moderate dynamics. The modal values of the Ca and Mg concentrations in waters sampled from caves in the vadose zone within karstic catchment of Prądnik river amount to Ca 90.0 and Mg 0.2 mg/l and in waters in phreatic zone from the whole Cracow Upland to Ca 96.0 and Mg 2.4 mg/l. Dissolution of the Upper Jurassic carbonate rocks in the saturation zone along groundwater flow paths shows an increase in concentration of components vulnerable to leaching in water from watershed area according to regional flow direction (from east towards the north east) and from tectonic horsts towards tectonic troughs. Within the phreatic zone calculated saturation indexes (SI) resulting from hydrogeochemical speciation modeling vary for carbonate minerals (particularly calcite) around the values close to equilibrium state, which is characteristic for diffuse flow system. The most prevalent forms of dissolved compounds are simple ion forms (Ca^{2+} and Mg^{2+}), in traces are sulphates ($CaSO_4^{o}$, $MgSO_4^{o}$) and acid carbonates (Ca, Mg)HCO_3^{+}.

REFERENCES

Bzowska G., Polonius A. & Różkowski J. (2008) Subordinate components of karstic rocky limestones and in fissure – karstic waters of the Cracow Jurassic. Mat. 43 Symp. Speleolog. Zamość 16–18.10.2009r. Sekcja Speleolog. Pol. Tow. Przyrod. w Krakowie, Inst. Ochrony Przyrody PAN. Kraków: 47–48.

Drever J.I. (1997) The Geochemistry of Natural Waters. *Surface and Groundwater Environments*. Prentice Hall, New Jersey 07458: 436.

Dżułyński S. & Żabiński W. (1954) Dark limestones of the Cracow Jurassic. *Acta Geol. Pol.*, 4, 1. Warszawa: 181–190.

Jeffries D.S. (1990) Snowpack storage of pollutants, release during melting, and impact on receiving waters. In Norton S.A., Lindberg S.E., Page A.L. (editors), *Acid Precipitation*. Volume 4: *Soils, Aquatic Processes, and Lake Acidification*. Springer-Verlag, New York, 107–132.

Johnson D.W. & Lindberg E. (1989) Acidic deposition on Walker Branch Watershed. In Adriano D.C., Havas M. (editors). *Acidic Precipitation*, 1, *Case Studies*. Springer-Verlag, New York, 1–38.

Jóźwiak A. & Kowalczewska G. (1984) Explanations to Hydrogeological map of Poland 1:200 000 Kraków sheet. *Wyd. Geol.* Warszawa.

Kulma R., Motyka J. & Rajpolt B. (1991) The chemical composition of groundwater flowing into the 'Zakrzówek' quarry in Cracow. *Gospodarka Surowcami Mineralnymi*, 7(1), 223–237.

Lis J. & Pasieczna A. (1995) *Geochemical Atlas of Upper Silesia 1:200000*. PIG. Warszawa.

Paczyński B. (1995) *Hydrogeological Atlas of Poland*. 1: 500 000. Part 2, *Resources, quality and protection of fresh groundwaters*. PIG. Warszawa.

Parkhurst D.L. & Appelo C.A. (1999) *User's Guide to Phreeqc (Version 2)–A Computer Program for Speciation, Batch-Reaction, One-Dimensional Transport and Inverse Geochemical Calculations*. U.S. Department of the Interior.

Różkowski A., Chmura A. & Siemiński A. (1997) Usable groundwaters in the Upper Silesian Coal Basin and its margin. *Prace PIG* CLIX. Warszawa.

Różkowski J. (2006) Groundwaters of carbonate formations in the southern part of Jura Krakowsko – Częstochowska and problems with their protection. Wyd. Uniw. Śląskiego. Katowice.

Różkowski J., Bzowska G., Polonius A. & Okoń D. (2009) Notes on the composition of the mineral deposits of karst spring niches and Upper Jurassic limestone massif of the Częstochowa Upland. Mat. 43 Symp. Speleolog. Zamość 16–18.10.2009r. Sekcja Speleolog. Pol. Tow. Przyrod. w Krakowie, Kraków, 76–78.

Różkowski J., Kowalczyk A., Rubin K. & Wróbel J. (2005) Groundwater circulation balance, renewal and resources in the Cracow Jurassic karstic aquifer in light of modelling study. *Kras i Speleologia*. Uniwersytet Śląski. 11(20), 187–199.

Różkowski K. (2008) Evolution of water chemistry in the vadose zone of the Upper Jurassic limestones of the southern part of Cracow – Częstochowa Upland. Ph.D. dissertation. Arch. AGH. Cracow.

Witczak S. & Adamczyk A. (1995) Catalog of selected physical and chemical indicators of groundwater pollution and the methods of their determining. T II. Biblioteka Monitoringu Środowiska. Warszawa.

Chapter 5

Groundwater calcium and magnesium content in various lithological types of aquifers in Slovenia

Kim Mezga & Janko Urbanc

Geological Survey of Slovenia, Department of Hydrogeology, Ljubljana, Slovenia

ABSTRACT

A 3-year geochemical study of groundwater has been undertaken in the main aquifers of Slovenia. Sampling locations include springs, groundwater observation boreholes, piezometers and wells, and surface waters. The chemical composition of Ca^{2+}, Mg^{2+} content and their molar ratios in groundwater were studied in 175 water samples. The objective was to evaluate groundwater Ca^{2+} and Mg^{2+} content in various lithological rock types in the aquifers. Based on the Ca^{2+}/Mg^{2+} molar ratio in groundwater it was possible to estimate the prevailing carbonate rock in the recharge area, and by calculating the saturation indexes of calcite and dolomite to determine the chemical equilibrium status between carbonate minerals and groundwater.

5.1 INTRODUCTION

5.1.1 The aim of the study

The aim of the study was to determine the characteristics of Ca^{2+} and Mg^{2+} in Slovenian groundwaters. The groundwater chemistry data from 87 different sampling locations was processed in order to determine the relationship between Ca^{2+} and Mg^{2+} concentrations and their molar ratio in groundwater in various different aquifer lithologies.

5.1.2 Description of study area

The Slovenian territory has been the site of complex geological processes from Palaeozoic to present, as shown through its geological structure and diverse lithology (Figure 5.1). Different rock types are present: sedimentary, igneous and metamorphic rocks. The most abundant are sedimentary rocks that underlie around 93% of Slovene territory. Among the sedimentary rocks the Mesozoic to Paleocene carbonate rocks (limestone and dolomite) are representative of the study area, especially for the southern and north-western part of the country (Javornik, 1989; Buser, 2010; Bavec & Pohar, 2009). Karst covers almost half of the territory of the country (Gams, 1974). Paleogene flysch rocks underlie large parts of south-western Slovenia while in the eastern and central parts of Slovenia Neogene clastic sedimentary rocks (sandstone, marl and siltstone) prevail. Unconsolidated Quaternary sediments (around 10%) cover basins in the central part to the north-eastern part of Slovenia. Igneous rocks (around 3%) make up a smaller parts of Slovenia, mainly in the north-eastern and

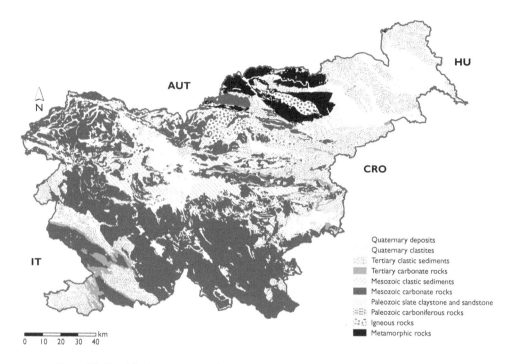

Figure 5.1 Simplified overview geological map of Slovenia (adapted after Buser, 2010).

northern part of the country. A smaller proportion of igneous rocks can also be found in central Slovenia where igneous rocks of mid-Triassic and Pliocene are present. The oldest rocks are metamorphic rocks of Palaeozoic age (around 4%), and they are located in north-eastern and northern Slovenia. On the north-northeastern area low grade metamorphic rocks of young Palaeozoic and high temperature metamorphic rocks are found (Javornik, 1989; Buser, 2010; Bavec & Pohar, 2009).

The greatest part of the Slovene territory is characterised by aquifers which are categorised as intergranular, fissured and karstic aquifers (Brenčič, 2009). In sandstone, shale, igneous and metamorphic rocks less significant aquifers of local and limited groundwater quality are present (Prestor *et al.*, 2001).

5.2   MATERIALS AND METHODS

5.2.1   Groundwater sampling strategy

A 3-year study (2009–2011) has been undertaken on 87 sampling locations across Slovenia in order to get a detailed insight into groundwater chemical status. The sampling locations (Figure 5.2) were evenly distributed over the entire Slovenian territory and are considered representative for the main aquifer types, defined by aquifer lithology. Sampling locations consist of springs (51), piezometers or wells (5), private wells (4), public water supply wells (18), and surface waters (9). Surface waters were

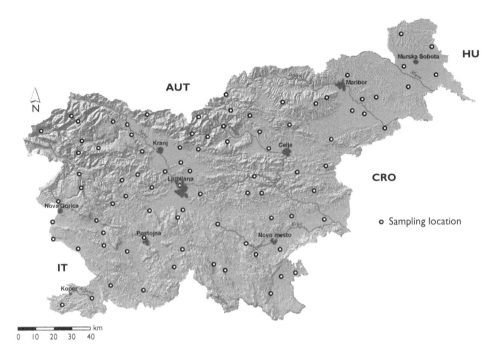

Figure 5.2 Distribution of sampling locations.

sampled where the access to groundwater was not possible. Sampling was performed during low flow hydrological conditions dominated by base flow, and covered three annual cycles. Groundwater at each sampling location was sampled twice (in spring and in autumn).

5.2.2   Sampling preparation

Standard procedures for sampling, transport and storage of groundwater samples were in accordance with standards (SIST ISO 5667-11:1996, SIST ISO 5667-03:1996, SIST ISO 5667-6:1996). Field measurements of physical parameters (electrical conductivity, pH and water temperature) of the sampled waters were carried out with the WTW pH/Conductivity measuring instrument pH/Cond 340i SET. Groundwater levels were measured by groundwater level measurement dipper. Each water sample was sealed in 1 litre polyethylene bottle for major ion chemical analysis.

5.2.3   Analytical methods

Major cations and anions in water were analysed in the Laboratory of public utility Vodovod-Kanalizacija d.o.o. (Slovenia). Concentrations of Ca^{2+} and Mg^{2+} were determined with the SIST EN ISO 14911:2000 standard in the ion chromatograph by Metrohm.

The analytical uncertainty for Ca^{2+} is between 4–14% ($n = 171$) and for Mg^{2+} 4.5–19% ($n = 164$). In cases of Ca^{2+} and Mg^{2+} where concentrations were below the Limit Of Detection (LOD) analytical uncertainty calculations in 4 water samples for Ca^{2+} and 11 water samples for Mg^{2+} were excluded.

5.3   RESULTS AND DISCUSSION

5.3.1   Determination of sampling locations recharge areas

Recharge areas were determined for every sampling location according to the aquifer type. If the aquifer is karstic or fissured the lithological and hydrogeological structure, the topology of terrain, active water protection areas, past tracer tests results, past hydro-contours, orographic watersheds and borders of groundwater bodies and aquifer systems were examined in detail. For groundwater sampled in intergranular aquifers the main hydrogeological characteristics of the aquifer, i.e. groundwater flow velocity and groundwater direction, were assessed. Additional information was provided by past hydrogeological modelling exercises for which the outer border of the recharge area was given by the 1-year isochrone. The recharge areas of sampling locations in aquifers with minor groundwater reservoirs, i.e. poorly permeable or impermeable rocks, were determined mostly from the terrain topography.

Mean altitudes of sampling location recharge areas were calculated in ArcGIS Version 9.2 as spatial analyses based on raster layers of the Slovenian Digital Elevation Model with a cell size of 12.5 × 12.5 m (Survey and Mapping Administration, 2000). Lithological composition of recharge areas was determined on the basis of the Geological map of Slovenia 1:250,000 (Buser, 2010). Since the lithology of Slovenia is very heterogenic the simplification was performed where rocks were divided into three major rock types and ten subgroups (Table 5.1).

5.3.2   Groundwater chemical composition

The statistical parameters such as minimum, maximum, mean, median, and standard deviation of Ca^{2+}, Mg^{2+} and Ca^{2+}/Mg^{2+} molar ratio in groundwater are presented in Table 5.2.

Table 5.1 Simplified lithological classification of typical Slovenian rock types.

Basic group	Description
Clastic sedimentary rocks	flysch rocks
	clay
	gravel and sand
	gravel, sand and clay
	shale and sandstone
Carbonate rocks	limestone prevailing
	dolomite prevailing
	carbonates with clastics
Igneous and metamorphic rocks	igneous rocks
	metamorphic rocks

Box and whisker plots for groundwater Ca^{2+} and Mg^{2+} are presented in Figure 5.3. Positive outliers in Ca^{2+} are found in groundwater sampled in clastic sedimentary rocks.

Concentrations for Ca^{2+}, Mg^{2+}, and Ca^{2+}/Mg^{2+} molar ratio in groundwater were tested for normality with a graphical analysis shown in the histograms (Figures 5.4

Table 5.2 Descriptive statistics for groundwater chemical composition ($n = 175$).

	Min	Max	μ	Md	s
Ca^{2+} [mg/l]	2.0	152.0	59.5	58.0	29.7
Mg^{2+} [mg/l]	1.0	42.0	12.9	9.5	10.1
Ca^{2+}/Mg^{2+} molar ratio	1.0	46.0	4.5	3.1	5.2

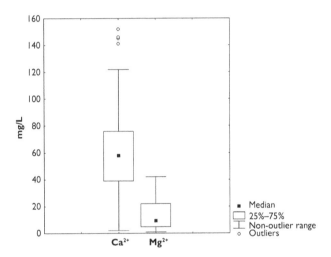

Figure 5.3 Box and whisker plots for Ca^{2+} and Mg^{2+} in sampled groundwater.

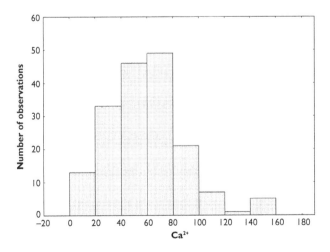

Figure 5.4 Histogram for Ca^{2+} content in groundwater ($n = 175$).

to 5.6) where Ca^{2+} in groundwater showed normal distribution and Mg^{2+} and Ca^{2+}/Mg^{2+} molar ratio non-normal (test of normality for skewness and kurtosis, Kolmogorov–Smirnov and Shapiro–Wilk tests).

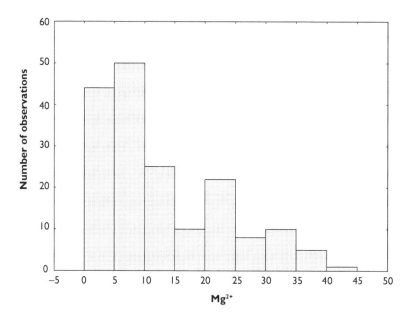

Figure 5.5 Histogram for Mg^{2+} content in groundwater ($n = 175$).

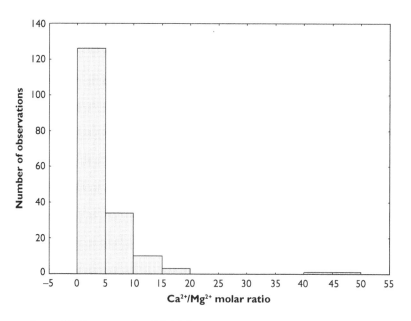

Figure 5.6 Histogram for Ca^{2+}/Mg^{2+} molar ratio in groundwater ($n = 175$).

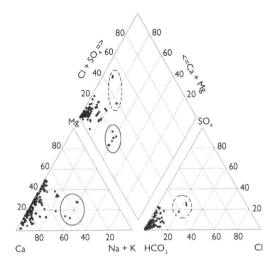

Figure 5.7 Piper diagram showing the composition of groundwater in the study area.

5.3.3 Hydrochemical classification and characterisation of water samples

Dominant dissolved species were plotted (program AquaChem®5.1) in a hydrochemical trilinear Piper diagram (Figure 5.7). The plot shows that most of the groundwater samples belong to calcium-magnesium rich water with a dominant Ca^{2+}–Mg^{2+}–HCO_3^- water type, and a secondary Ca^{2+}–HCO_3^- water type. Groundwater Ca^{2+}, Mg^{2+} and HCO_3^- result from dissolution of carbonate rocks dominating in the recharge area of sampling locations. The third and fourth largest groups belong to Na^+–Ca^{2+}–Mg^{2+}–HCO_3^- and Ca^{2+}–Na^+–HCO_3^- water types, which contain earth alkali elements and weak acidic anions but also more alkali metals. The minority of the water samples belong to Ca^{2+}–Mg^{2+}–HCO_3^-–NO_3^-, Ca^{2+}–HCO_3^-–SO_4^{2-}, and Ca^{2+}–Mg^{2+}–HCO_3^-–SO_4^{2-} water types, which contain more strong acidic anions. It is believed that the source of NO_3^- in groundwater comes from agriculture and urbanisation. The source of SO_4^{2-} in groundwater in high mountain region is likely to be of natural origin from carbonate rocks containing gypsum (Brenčič & Polting, 2008; Vidrih, 2006).

Water samples indicated by a dashed line belong to intergranular aquifers in the north-eastern part of Slovenia where groundwater contains elevated values of Cl^- and SO_4^{2-} but less HCO_3^-. The sources of first two parameters in groundwater are likely to be of anthropogenic origin. Water samples bounded by the solid line have their recharge areas in metamorphic rocks and in clay, and contain more alkali metals. The observed trend could be a consequence of cation-exchange in clay minerals or a process of plagioclase mineral (albite) weathering.

Table 5.3 presents the types and number of sampled groundwaters with regard to the lithology of the recharge area ($n = 175$).

Table 5.3 Types of water with regard to the lithology of the recharge area.

Rock type group	Geology	Water type						
		$Ca^{2+}–HCO_3^-$	$Ca^{2+}–HCO_3^-–SO_4^{2-}$	$Ca^{2+}–Mg^{2+}–HCO_3^-$	$Ca^{2+}–Mg^{2+}–HCO_3^-–NO_3^-$	$Ca^{2+}–Mg^{2+}–HCO_3^-–SO_4^{2-}$	$Ca^{2+}–Na^+–HCO_3^-$	$Na^+–Ca^{2+}–Mg^{2+}–HCO_3^-$
Clastic sedimentary rocks	Flysch rocks	7						
	Clay							
	Gravel, sand	5		18			2	
	Gravel, sand and clay	2		9			1	
	Shale and sandstone		2	5				
Carbonate rocks	Limestone prevailing	52		21	1			
	Dolomite prevailing	2		28				
	Carbonates with clastics	2		4				
Igneous and metamorphic rocks	Igneous rocks	6						2
	Metamorphic rocks	2						2

5.3.4   Saturation indexes of calcite and dolomite in groundwater

Saturation Indexes (SI) for calcite and dolomite (Figure 5.8) were calculated using the program PHREEQC (Parkhurst & Appelo, 1999) interfaced with AquaChem®5.1 program. Most Slovenian groundwaters in carbonate rocks are supersaturated with respect to calcite and dolomite (SI mostly up to 1). Groundwater is undersaturated with respect to calcite in 31 (17.7%) water samples and with respect to dolomite in 81 (46.3%), which indicates the lack of carbonate minerals in groundwater some recharge area.

5.3.5   Groundwater Ca^{2+}/Mg^{2+} molar ratio

A Ca^{2+}/Mg^{2+} molar ratio (Figure 5.9) >1 prevails in all the sampled groundwater. Lower Ca^{2+}/Mg^{2+} molar ratios (<4) indicate a groundwater origin in dolomite and mixed dolomite with limestone, where the water types are Ca^{2+}–Mg^{2+}–HCO_3^- and Na^+–Ca^{2+}–Mg^{2+}–HCO_3. Groundwater with higher molar ratio (>4) is mostly of Ca^{2+}–HCO_3^- water type indicating a limestone prevalence in the recharge area. Two water samples from one location were excluded due to high molar ratio (>40) indicating very pure limestone in the recharge area.

5.3.6   Groundwater Ca^{2+}, Mg^{2+} and Ca^{2+}/Mg^{2+} molar ratio in various lithological units

Table 5.4 presents minimum, maximum, mean and median values of groundwater Ca^{2+} and Mg^{2+} concentrations, and their molar ratios in various types of lithological units included in this study.

5.3.6.1 Ca^{2+} in groundwater

Basic descriptive statistics for Ca^{2+} in groundwater for all three major rock types (Table 5.5) and the box and whisker plots (Figure 5.10) for groundwater Ca^{2+} are presented below.

Groundwater content of Ca^{2+} found in recharge areas in Quaternary sand, gravel and clay, and in flysch rocks are statistically higher compared to groundwater in clays,

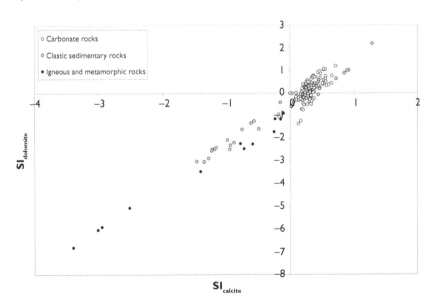

Figure 5.8 Scatter plot of calcite saturation index ($SI_{calcite}$) versus dolomite saturation index ($SI_{dolomite}$).

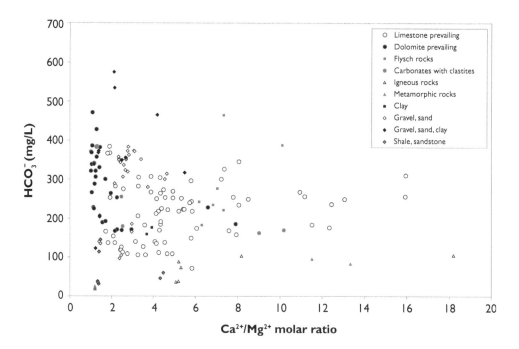

Figure 5.9 Ca^{2+}/Mg^{2+} molar ratio versus HCO_3^- (two groundwater samples excluded).

shale and sandstone, and igneous and metamorphic rocks, which have the lowest content of groundwater Ca^{2+} ($p < 0.05$) (Figure 5.11).

A Flysch rocks

Median Ca^{2+} groundwater content in flysch rocks recharge area is 70.00 mg/l and is quite high. The origin of groundwater Ca^{2+} is in sandstone and marl, in which the carbonate cement binds quartz grains in sandstone (Pavlovec, 1977, 1980). Marls are softer and enriched with $CaCO_3$.

B Clay

In clay a lower groundwater Ca^{2+} content is found (36.50 mg/l). Clays are mostly impermeable and the marine clay termed *sivica* is composed of phyllosilicates (illite/ muscovite, chlorite and montmorillonite) and carbonates (Kralj & Mišič, 2003).

C Sand and gravel

High median values are found in recharge area in sand and gravel (89.50 mg/l). The outlier in the group shows an absence of groundwater Ca^{2+} and HCO_3^- due to the alluvial deposits which have mostly quartz and minerals of igneous and metamorphic rocks in their recharge area.

Table 5.4 The Ca²⁺ and Mg²⁺ concentration and Ca²⁺/Mg²⁺ molar ratio in various lithological units.

Basic group	Description	No. of samples	Ca^{2+} [mg/l]				Mg^{2+} [mg/l]				Ca^{2+}/Mg^{2+} molar ratio			
			Min	Max	μ	Md	Min	Max	μ	Md	Min	Max	μ	Md
Clastic sedimentary rocks	Flysch rocks	7	53.0	145.0	87.5	72.0	5.1	12.0	7.1	6.3	6.1	10.1	7.3	7.0
	Clay	2	36.0	37.0	36.5	36.5	5.7	5.9	5.8	5.8	3.7	3.9	3.8	3.8
	Gravel and sand	26	28.0	122.0	84.7	89.5	11.0	26.0	18.8	22.0	1.4	4.6	2.7	2.7
	Gravel, sand and clay	11	20.0	152.0	90.3	95.0	9.8	42.0	24.5	22.0	1.2	5.5	2.3	2.1
	Shale and sandstone	7	6.4	29.0	14.9	15.0	2.1	12.0	5.0	3.4	1.3	4.4	2.2	1.4
Carbonate rocks	Limestone prevailing	74	25.0	92.0	57.7	61.5	1.0	27.0	7.9	7.3	1.7	46.0	6.3	4.4
	Dolomite prevailing	30	39.0	76.0	55.0	55.0	4.3	36.0	23.7	25.5	1.0	7.9	1.8	1.2
	Carbonates with clastics	6	46.0	73.0	56.0	49.0	2.8	34.0	16.0	11.5	1.3	10.1	4.4	2.5
Igneous and metamorphic rocks	Igneous rocks	8	2.0	31.0	16.1	15.3	1.0	2.9	1.5	1.0	1.2	18.1	6.2	5.2
	Metamorphic rocks	4	2.00	22.0	11.2	10.5	1.0	1.0	1.0	1.0	1.2	13.3	6.8	6.3

Table 5.5 Basic descriptive statistics for Ca²⁺ in groundwater for all three major rock types [mg/l].

	n	*μ*	*Md*	*Min*	*Max*	*s*
Carbonate rocks	110	56.8	57.0	25.0	92.0	17.6
Clastic sedimentary rocks	53	75.2	79.0	6.4	152.0	39.4
Igneous and metamorphic rocks	12	14.5	13.8	2.0	31.0	11.4

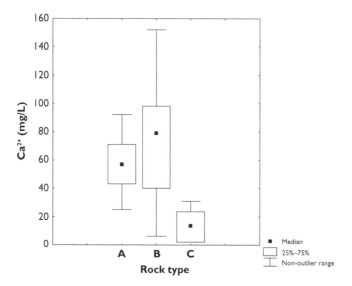

Figure 5.10 Box and whisker plots for Ca²⁺ in groundwater for all three major rock major rock (A = carbonate rocks, B = clastic sedimentary rocks, C = igneous and metamorphic rocks).

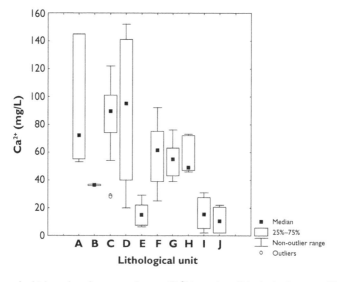

Figure 5.11 Box and whisker plots for groundwater Ca²⁺ in various lithological groups (A = flysch rocks, B = clay, C = gravel and sand, D = gravel, sand and clay, E = shale and sandstone, F = limestone prevailing, G = dolomite prevailing, H = carbonates with clastics, I = igneous rocks, J = metamorphic rocks).

Figure 5.12 presents the comparison between Ca^{2+} and HCO_3^- measured in groundwater in intergranular aquifers (gravel, sand and clay). Sampling locations in the encircled area belong to groundwater sampled in mostly non-carbonate alluvial deposits of the Mura River (north-eastern part) and deviate from other groundwaters by their lower Ca^{2+} and HCO_3^- content. Most alluvial deposits belong to Quaternary sediments of carbonate and silicate origin (Bavec & Pohar, 2009; Slovenian Environment Agency, 2009). In the Mura depression (north-eastern area), the gravel is characterised by pebbles of quartz, igneous and metamorphic rocks, and to a lower extend, by pebbles of older sedimentary rocks originating from the Central Alps (Markič, 2009).

D Gravel, sand and clay

Median groundwater Ca^{2+} content in recharge area of gravel, sand and clay is the highest (95.00 mg/l). Conglomerate pebbles and grains are of metamorphic, carbonate and igneous rocks, and in sandstones the carbonate cement prevails. In marl a lot of mica could be found (Mioč & Žnidarčič, 1989). In deep aquifers (south-eastern area) the source of groundwater Ca^{2+} in gravel, sand and clay is in carbonates and phyllosilicates (Kralj, 2003).

E Shale and sandstone

The content of groundwater Ca^{2+} in shale and sandstone is the smallest (15.00 mg/l) due to mostly noncarbonate clastic sediments (Skaberne *et al.*, 2009).

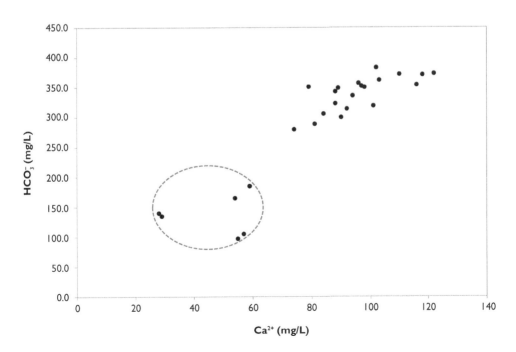

Figure 5.12 Groundwater Ca^{2+} versus HCO_3^-.

F Limestone prevailing, G Dolomite prevailing and H Carbonates with clastics

Median values of Ca^{2+} in limestone recharge areas is 61.50 mg/l, in dolomite recharge area 55.00 mg/l and in carbonates with clastics recharge area 49.00 mg/l. All these rocks are rich in calcite and dolomite. Dissolution of carbonates mainly depends on the presence of CO_2 in solution which enables percolating water to dissolve carbonate minerals from rocks and soils, thus adding to the total HCO_3^- (Singhal & Gupta, 1999).

Figure 5.13 presents the relation between HCO_3^- in groundwater and the mean altitude of the recharge area of the sampled groundwater. Water at higher altitudes has less dissolved bicarbonate (HCO_3^-) compared to groundwater at lower altitudes. This is due to the thin soil layer which cannot enrich the atmospheric CO_2 (P_{CO_2}) in groundwater and produce carbonic acid which dissolves carbonate minerals and produces HCO_3^-.

I Igneous rocks and J Metamorphic rocks

Low Ca^{2+} content in groundwater was observed in igneous (15.30 mg/l) and metamorphic rocks (10.50 mg/l), because minerals in igneous granodiorite and metamorphic biotite and muscovite gneiss are not rich in calcium. Igneous and metamorphic rocks are relatively impermeable for water so that rainfall runs off into surface streams and

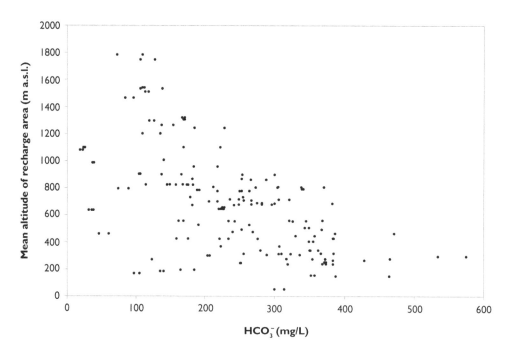

Figure 5.13 Scatter plot of HCO_3^- in groundwater versus mean altitude of recharge area.

enters the aquifer in places where rocks are sufficiently fractured. Water type of igneous and metamorphic rocks in the recharge area suggest an increase in dissolution of Na^+ ions which is attributed to cation exchange with aquifer materials (see Piper diagram). The groundwater Ca^+ values in both cases were below LOD.

5.3.6.2 Mg^{2+} in groundwater

Basic descriptive statistics for Mg^{2+} in groundwater for all three major rock types are presented in Table 5.6, and the box and whisker plot for Mg^{2+} in groundwater in Figure 5.14.

In recharge areas in gravel, sand and clay, dolomite and carbonates with clastics, the groundwater Mg^{2+} content was significantly higher compared to groundwater Mg^{2+} content in other lithological units ($p < 0.05$) (Figure 5.15). Medium groundwater Mg^{2+} values were found in recharge areas in flysch rocks, clay, and sand and gravel, shale and sandstone, and limestone. The lowest groundwater Mg^{2+} contents were measured in recharge area in igneous and metamorphic rocks.

Table 5.6 Basic descriptive statistics for Mg^{2+} in groundwater for all three major rock types [mg/l].

	n	μ	Md	Min	Max	s
Carbonate rocks	110	12.6	8.7	1.0	36.0	10.1
Clastic sedimentary rocks	53	16.1	14.0	2.1	42.0	9.3
Igneous and metamorphic rocks	12	1.3	1.0	1.0	2.9	0.7

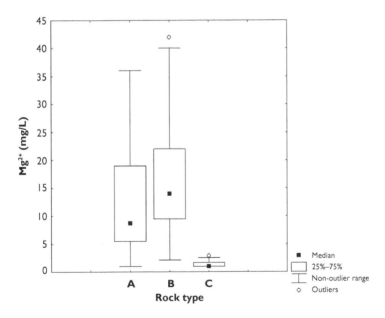

Figure 5.14 Box and whisker plots for Mg^{2+} in groundwater for all three major major rock (A = carbonate rocks, B = clastic sedimentary rocks, C = igneous and metamorphic rocks).

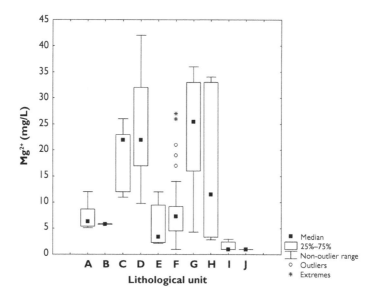

Figure 5.15 Box and whisker plots for Mg^{2+} in groundwater in various lithological groups (A = flysch rocks, B = clay, C = gravel and sand, D = gravel, sand and clay, E = shale and sandstone, F = limestone prevailing, G = dolomite prevailing, H = carbonates with clastics, I = igneous rocks, J = metamorphic rocks).

A Flysch rocks

Low values of Mg^{2+} in groundwater (6.30 mg/l) are found in recharge area of flysch rocks for which the source of magnesium is in carbonate cement of sandstone and marl (Pavlovec, 1977, 1980).

B Clay

Also in marine clay called *sivica* recharge area the content of groundwater Mg^{2+} is low (5.80 mg/l). Sources of magnesium are found in phyllosilicates (illite/muscovite, chlorite and montmorillonite) and carbonates (Kralj & Mišič, 2003).

C Sand and gravel

High groundwater Mg^{2+} content is found in recharge area with sand and gravel (22.00 mg/l) which are mostly of carbonate origin except in north-eastern part of the country (Bavec & Pohar, 2009).

D Gravel, sand and clay

The content of groundwater Mg^{2+} in gravel, sand and clay recharge areas is high (22.00 mg/l). The source of groundwater Mg^{2+} are in the magnesium-rich minerals (dolomites, amphiboles and phyllosilicates) in recharge area of carbonate, metamorphic and igneous rocks (Mioč & Žnidarčič, 1989; Kralj, 2003).

E Shale and sandstone

In shale and sandstone recharge area the groundwater Mg^{2+} contents is low (3.40 mg/l). The source is likely to be phyllosilicate and carbonate minerals (Skaberne *et al.*, 2009).

*F Limestone prevailing, G Dolomite prevailing, and
H Carbonates with clastics*

High groundwater Mg^{2+} is found in carbonate rock recharge area: limestone prevailing (7.30 mg/l), dolomites prevailing (25.50 mg/l), and carbonates with clastics (11.50 mg/l). The source of groundwater Mg^{2+} is in the magnesium-rich minerals (in dolomites). In the limestone prevailing recharge areas significant amounts of groundwater Mg^{2+} can occur. Outliers in the limestones are where the prevailing recharge area is in pure limestones. The wide range in groundwater Mg^{2+} in carbonates with clastics recharge areas is due to heterogenic lithological structure within rock sequence. These rocks are mostly of carbonate origin and contain magnesium-rich minerals, which solubility depends on the presence of H_2CO_3 in the water.

I Igneous rocks and J Metamorphic rocks

The lowest groundwater Mg^{2+} contents were measured in igneous (1.00 mg/l) and metamorphic rocks (1.00 mg/l), where only three water samples had a level of Mg^{2+} above the LOD, and nine water samples were below the LOD. Although magnesium occurs in igneous and metamorphic rocks in the form of insoluble silicates, weathering could break them down into more soluble carbonates, clay minerals and silica, but it depends on the presence of H_2CO_3 in the water (Karanth, 1987).

5.3.6.3 Ca^{2+}/Mg^{2+} molar ratio in groundwater

Basic descriptive statistics for Ca^{2+}/Mg^{2+} molar ratio in groundwater for three major rock types are presented in Table 5.7, and box and whisker plots for Ca^{2+}/Mg^{2+} in groundwater in Figure 5.16.

Figure 5.17 presents the box and whisker plots for Ca^{2+}/Mg^{2+} molar ratio in groundwater in various lithological groups.

The highest Ca^{2+}/Mg^{2+} molar ratios in groundwater are observed in igneous and metamorphic rock recharge areas, with medium values in flysch rocks, carbonates with clastics, and clay recharge areas. The lowest values of groundwater molar ratio were observed in sand and gravel, gravel, sand and clay, shale and sandstone, and dolomite recharge areas.

Table 5.7 Basic descriptive statistics for Ca^{2+}/Mg^{2+} molar ratio in groundwater for all three major rock types.

	n	μ	Md	Min	Max	S
Carbonate rocks	110	5.0	3.7	1.0	46.0	6.1
Clastic sedimentary rocks	53	3.2	2.7	1.2	10.1	1.9
Igneous and metamorphic rocks	12	6.4	5.2	1.2	18.1	5.4

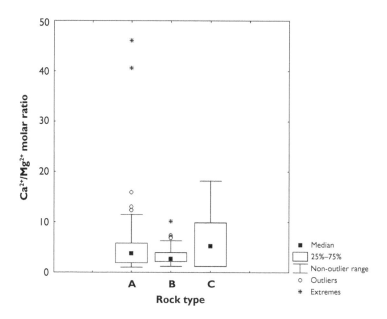

Figure 5.16 Box and whisker plots for Ca^{2+}/Mg^{2+} molar ratio in groundwater for all three major rock (A = carbonate rocks, B = clastic sedimentary rocks, C = igneous and metamorphic rocks).

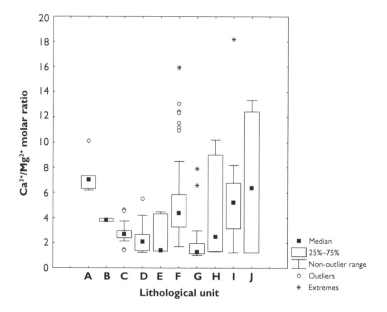

Figure 5.17 Box and whisker plot for Ca^{2+}/Mg^{2+} molar ratio in groundwater in various lithological groups (2 groundwater samples excluded).

The wide range of the Ca^{2+}/Mg^{2+} molar ratio in metamorphic rocks (J) is due to Ca^{2+} and Mg^{2+} content which were below LOD in nine water samples. In limestone recharge areas (F) a few outliers and extreme values occur due to pure limestone recharge area. The outliers in sand and gravel (C) are observed in sampling location in the alluvial deposits (north-east part) with lower carbonate content. Outlier with the highest molar ratio in gravel and sand (C) was found in the carbonate (limestone) recharge area of Soča River (western part). In the dolomite (G) recharge area one sampling location appears as an outlier due to the highest molar ratio, suggesting the limestone prevalence over dolomite in the recharge area. In igneous rocks (I) the outlier is groundwater with Mg^{2+} below LOD.

5.3.7 Spatial distribution of measured parameters

The results of chemical analysis of sampled groundwaters were used to produce hydrochemical thematic maps of Slovenia (program ArcGIS Version 9.2). Mean values of groundwater Ca^{2+}, Mg^{2+} and Ca^{2+}/Mg^{2+} molar ratio are presented in the maps (Figures 5.18 to 5.20).

5.3.8 Practical aspects of groundwater mineralisation

Until recently, in Slovenia the German water hardness scale was usually used, but in order to make the scale of water hardness internationally comparable, a new scale

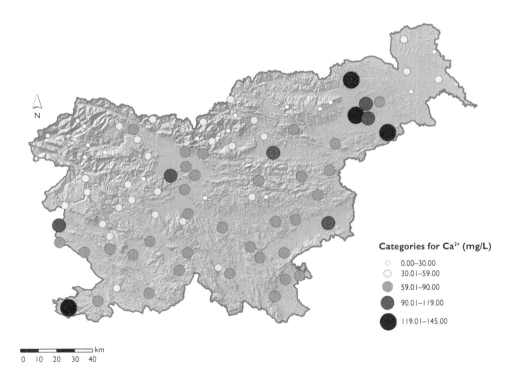

Figure 5.18 Spatial distribution of Ca^{2+} concentration in groundwater.

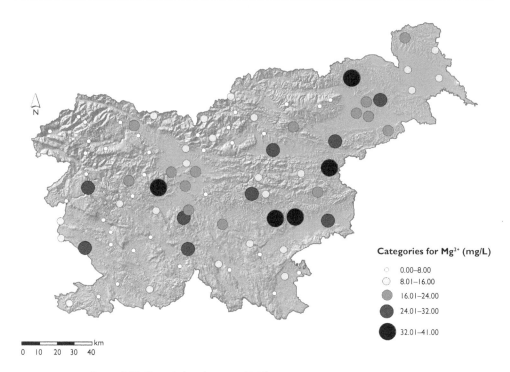

Figure 5.19 Spatial distribution of Mg²⁺ concentration in groundwater.

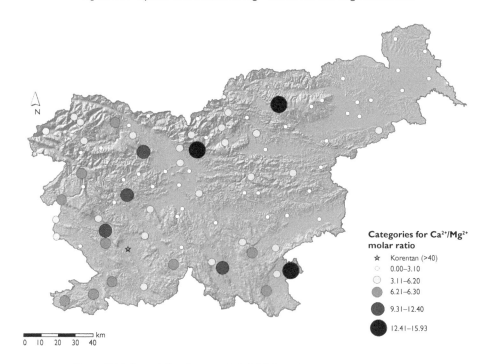

Figure 5.20 Spatial distribution of Ca²⁺/Mg²⁺ molar ratio in groundwater.

was introduced (Table 5.8) where calcium carbonate ($CaCO_3$) [mmol/l] is used as a unit for classification (DVGW, 2013).

In Slovenia medium hard groundwater prevails (Figures 5.21 and 5.22). The minimum and maximum water hardness are 0.09 mmol/l and 5.26 mmol/l as $CaCO_3$, with a mean value of 2.02 mmol/l and median 1.97 mmol/l, respectively. Soft water is mostly observed in recharge areas of metamorphic and igneous rocks, clay, carbonates with clastics, gravel, sand (and clay) sediments, shale and sandstone, and in high mountains carbonates. Hard water is observed in carbonates, gravel, sand (and clay) sediments and flysch rocks (north-east, central, and south-west area).

Groundwater from the high mountain regions (mostly carbonate rocks in recharge areas) is soft due to a very thin (or non-existent) layer of soil and very little vegetation (low level of respiration of soil organisms and the decay of organic matter), low temperatures, and consequently low P_{CO_2}, surface runoff during snow melting, and also because most sampling locations in high mountains are springs, and

Table 5.8 A new 3-point water hardness scale $CaCO_3$ [mmol/l] and its corresponding values of German degrees [°dH].

Water hardness	$CaCO_3$ [mmol/l]	German hardness [°dH]
Soft	<1.5	<8.4
Medium hard	1.5–2.5	8.4–14
Hard	>2.5	>14

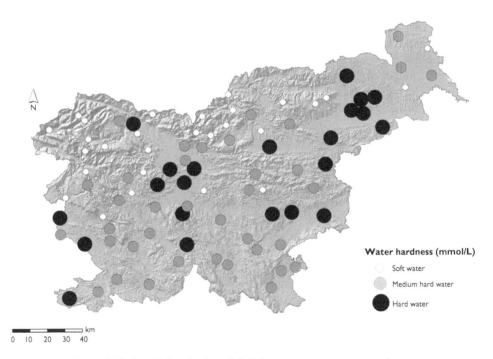

Figure 5.21 Spatial distribution of $CaCO_3$ concentration in groundwater.

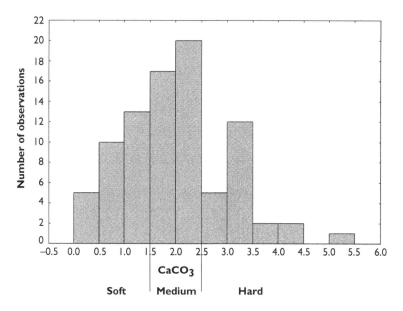

Figure 5.22 Histogram for CaCO$_3$ (mmol/L) content in groundwater (n = 175).

discharge responds primarily to snowmelt runoff. Soft groundwater is also observed in recharge areas of igneous and metamorphic rocks which contain very little dissolved Ca^{2+} and Mg^{2+}. Medium hard groundwater is found across the whole country except in the high mountain regions, and hard groundwater occurs in recharge areas of carbonate alluvial deposits and carbonate rocks. Medium hard and hard groundwater contain more Ca^{2+}, Mg^{2+} due to calcium- and magnesium-rich minerals in aquifers rocks, higher air temperatures due to lower altitude of recharge areas, thicker soil layer, increased microbiological activity and consequently greater P$_{CO_2}$, which increases solubility of minerals.

5.4 CONCLUSIONS

The hydrochemical investigation of Ca^{2+} and Mg^{2+} in 175 groundwater samples in various water-bearing formations of various aquifers in Slovenia reveal that groundwater mostly belongs to Ca^{2+}–HCO$_3^-$ and Ca^{2+}–Mg^{2+}–HCO$_3^-$ water type. Most Slovenian groundwaters in carbonate recharge areas are supersaturated with respect to calcite and dolomite (SI mostly up to 1) which is to be expected because a large part of the Slovenian territory is covered by carbonate rocks. The Ca^{2+}/Mg^{2+} molar ratios in sampled groundwater suggest the groundwater is in contact with either dolomites, limestones, or both. The amount of Ca^{2+} and Mg^{2+} in groundwater differs between lithological units in recharge areas due to the content and solubility of the prevailing minerals. Ca^{2+} in groundwater is derived mostly from calcium-rich minerals like calcite and gypsum in carbonate rocks. The source of groundwater Mg^{2+} is in magnesium-rich minerals such as dolomites, amphiboles and clay minerals. In Slovenia medium hard groundwater prevails with medium high levels of dissolved groundwater Ca^{2+} and Mg^{2+} in the form of carbonates which is a consequence of natural weathering of sedimentary rock in the study area.

ACKNOWLEDGEMENTS

The authors would like to thank the Slovenian Research Agency (ARRS) for the financial support obtained during the three-year project 'Natural hydrochemical background and dynamics of groundwater in Slovenia', which was implemented at the Geological Survey of Slovenia.

REFERENCES

Bavec M. & Pohar V. (2009) Quaternary (Chapter 6.5). In Pleničar M., Ogorelec B., Novak M. (Editors) Geologija Slovenije (The geology of Slovenia). Geological Survey of Slovenia, Ljubljana. pp. 441–464.

Brenčič M. (2009) Shallow groundwater (Chapter 9.1). In Pleničar M., Ogorelec B., Novak M. (Editors) Geologija Slovenije (The geology of Slovenia). Geological Survey of Slovenia, Ljubljana. pp. 543–553.

Brenčič M. & Poltnig W. (2008) Groundwater Karavanke: hidden wealth. Geological Survey of Slovenia, Ljubljana & Joanneum Research Forschungsgesellschaft, Graz, 143 p.

Buser S. (2010) Geologic map of Slovenia 1:250,000 scale. Geological Survey of Slovenia, Ljubljana.

DVGW (2013) (Deutscher Verein des Gas- and Wasserfaches e.V) Neue Härtebereiche für Trinkwasser. Available from [25.09.2013]: http://www.dvgw.de/wasser/informationen-fuer-verbraucher/wasserhaerte/

Gams I. (1974) Kras. Zgodovinski, naravoslovni in geografski oris (Karst: historical, natural and geographic features). Slovenska matica, Ljubljana. 358 p.

Javornik M. (Editor) (1989) Enciklopedija Slovenije (Encyclopedia of Slovenia). Mladinska knjiga, Ljubljana. pp. 194–203.

Karanth K.R. (1987) Ground water assessment, development, and management. Tata McGraw-Hill Publishing Company, New York. 720 p.

Kralj P. & Mišič M. (2003) Chemical composition of Kiscellian silty sediment (sivica) from the Trobni Dol area, Eastern Slovenia. Geologija 46/1, Ljubljana, 113–116. Available from [09.02.2013]: http://www.geologija-revija.si/dokument.aspx?id=579, 113-116

Kralj P. (2003) Geochemistry of Upper Pliocene silty and sandy sediments from the well Mt-7, Moravci Spa, North-Eastern Slovenia. Geologija 46/1, Ljubljana, pp. 117–122. Available from [12.02.2013]: http://www.dlib.si/details/URN:NBN:SI:doc-380IGI3E/

Markič M. (2009) Pliocene and Plio-Quaternary (Chapter 6.4). In Pleničar M., Ogorelec B., Novak M. (Editors) Geologija Slovenije (The geology of Slovenia). Geological Survey of Slovenia, Ljubljana. pp. 427–440.

Mioč P. & Žnidarčič M. (1989) Osnovna geološka karta SFRJ. 1:100 000. Tolmač za lista Maribor in Leibnitz: L 33-56, L 33-44. Beograd: Zvezni geološki zavod, 60 p.

Parkhurst D.L. & Appelo, C.A.J. (1999) User's guide to PHREEQC (version 2) – A computer program for speciation, batch-reaction, one-dimensional transport, and inverse geochemical calculations: U.S. Geological Survey Water-Resources Investigations Report 99-4259, 312 p.

Pavlovec R. (1977) Geološki sprehod po slovenski obali. Planinski vestnik 77(7), 401–408.

Pavlovec R. (1980) Preperevanje v flišnem peščenjaku. Proteus 43(3), 120.

Pleničar M., Ogorelec B. & Novak M. (Editors) (2009) Geologija Slovenije (The geology of Slovenia). Geological Survey of Slovenia Ljubljana. 612 p.

Prestor J., Janža M., Rikanovič R. & Strojan M. (2001) Dosegljivost, izkoristljivost in izkoriščenost podzemnih vodonosnikov (Accessibility, exploitability and utilization of underground aquifers). Geological Survey of Slovenia, Ljubljana.

Singhal B.B.S. & Gupta R.P. (1999) Applied Hydrogeology of Fractured Rocks. Kluwer, Dordrecht, 393 p.

Skaberne D., Ramovš D. & Ogorelec B. (2009) Middle and upper Permian (Chapter 4.5). In Pleničar M., Ogorelec B., Novak M. (Editors) Geologija Slovenije (The geology of Slovenia). Geological Survey of Slovenia, Ljubljana. pp. 137–154.

Slovenian Environmental Agency (2009) Kakovost podzemne vode v Sloveniji v letih 2007 in 2008.

Survey and Mapping Administration (2000) InSAR DMV 12,5 (digitalni model višin). (InSAR DEM 12.5 (digital elevation model).

Vidrih R. (2006) Minerali karavanškega predora. Scapolia Suppl.3. pp. 125–127. Available from [10.02.2013]: http://www.landesmuseum.at/pdf_frei_remote/Scopolia_Suppl_3_0125-0127.pdf.

Chapter 6

Ca and Mg in fractured and karstic aquifers of Slovenia

Timotej Verbovšek[1] *& Tjaša Kanduč*[2]

[1]*Department of Geology, Faculty of Natural Sciences and Engineering, University of Ljubljana, Ljubljana, Slovenia*
[2]*Institut Jožef Stefan, Ljubljana, Slovenia*

ABSTRACT

Slovenia's carbonates represent an ideal research polygon for the karst and fractured aquifers, as the limestone and dolomite sedimentation persisted almost continuously from the Upper Permian to the end of the Mesozoic. This process resulted in deposition of carbonate rocks which are several kilometers thick. A database of 430 boreholes in dolomite and limestone aquifers has been constructed, out of which 90 boreholes used for public water supply systems, were chosen for hydrogeochemical and isotopic analyses. Results indicate several differences between individual dolomite and limestone aquifers for most of analyzed geochemical parameters. Total alkalinity ranges from 2.1 mM to 8.6 mM. The study is focused on calcium and magnesium concentrations within aquifer and revealed, that the average value for Ca^{2+} in all aquifers is 63.8 mg/l (ranging from 22.7 to 313 mg/l), while for Mg^{2+} is 30.8 mg/l (ranging from 6.7–72.0 mg/l). Mg^{2+}/Ca^{2+} molar ratios lie mostly in the range 0.20 to 1.84, indicating weathering of pure dolomites; however large value spans occur for different aquifers, which can be explained by the composition and origin of dolomite rocks. The highest values of Mg^{2+}/Ca^{2+} ratio appear for late-diagenetic Cordevolian dolomites, while the lowest for relatively impure Lower Triassic dolomites. $\delta^{13}C_{DIC}$ range from −14.6 to −8.2‰ and reveal the contribution of organic matter and dissolution of carbonates within aquifer. According to mass balance calculation it was estimated that contribution of organic matter within aquifer ranged from 41.0 to 62.2% and dissolution of carbonates from 37.8 to 59.0‰.

6.1 INTRODUCTION

Carbonates are an important groundwater resource and are abundant in many countries. Despite their relative abundance, the research of karstic and fractured aquifers is still insufficient, mostly because of their complexity. Slovenia, being the homeland of karst, presents an ideal polygon for dolomite and limestone research, as more than a 7 km thick sequence of almost non-interrupted carbonate deposition had occurred. Within those, several different dolomite and limestone sequences can be distinguished based on their lithological, geochemical and other properties. The main reason that carbonates are inadequately investigated is their large heterogeneity (de Marsily *et al.*, 2005).

The high solubility of carbonate minerals compared with that of silicate minerals (e.g. feldspars) causes the weathering of carbonate minerals to dominate the geochemical signatures and the concentration of elements in water systems (Drever, 1997). Thermodynamic data for carbonate mineral dissolution indicate that dolomite has a progressively greater absolute and relative solubility to calcite at temperatures below 25°C (Langmuir, 1997). Calcite and dolomite dissolution occurs via carbonic acid (H_2CO_3). Dissolution of calcite and dolomite produces waters with a molar ratio of ($Ca^{2+} + Mg^{2+}$): $HCO_3^- = 1:2$ (Eqs. 1 and 2):

Calcite: $CaCO_3 + H_2CO_3 \leftrightarrow Ca^{2+} + 2HCO_3^-$ (6.1)

Dolomite: $Ca_xMg_{1-x}CO_3 + H_2CO_3 \leftrightarrow xCa^{2+} + (1-x) Mg^{2+} + 2HCO_3^-$ (6.2)

The weathering of dolomite in carbonate systems contributes the majority of Mg^{2+}, in which the Mg^{2+}/Ca^{2+} and Mg^{2+}/HCO_3^- molar ratios indicate the relative proportions of calcite and/or dolomite dissolution (Williams *et al.*, 2007; Zavadlav *et al.*, 2013). Dissolution of calcite produces waters with a Mg^{2+}/Ca^{2+} molar ratio of less than 0.1, 0.33 in the case of congruent dissolution of calcite and dolomite, and equal to 1 if only dolomite is dissolving (Szramek *et al.*, 2011).

Measurements of stable carbon isotopes, combined with other groundwater chemistry parameters, can be important tools to understand the geochemistry of ground water systems (Li *et al.*, 2005). Stable carbon isotopes are useful indicators of DIC (dissolved inorganic carbon) sources in groundwater systems and are used to assess the origin of DIC, which is the main species in carbonate environments. Concentrations of DIC and its stable carbon isotope ratios ($\delta^{13}C_{DIC}$) are governed by processes occurring in the aquifer system, and these vary seasonally. Changes in DIC concentrations result from carbon addition or removal from the DIC pool, while changes of $\delta^{13}C_{DIC}$ result from the fractionation accompanying transformation of carbon or from mixing of carbon from different sources. The major sources of carbon to aquifer DIC loads are dissolution of carbonate minerals, soil CO_2 derived from root respiration and from microbial decomposition of organic matter (Aucour *et al.*, 1999; Li *et al.*, 2005; Kanduč *et al.*, 2007). The major process removing DIC from aquifer systems is carbonate mineral precipitation (Atekwana and Krishnamurthy, 1998).

In the present work, 90 wells have been selected for the analysis of groundwater in these aquifers in several different lithologies. The aim of research was to study the subtle geochemical and isotopic differences among different dolomites (mostly) and limestones and to evaluate geochemical processes within investigated aquifers.

Slovenian legislation (Rules on Drinking Water, Official Gazette of Republic of Slovenia, no. 19/2004) does not put any limits on Ca and Mg concentration in drinking water, so it is still an open question whether such concentrations put any health hazard to humans. The purpose of this paper is to investigate the concentrations of Ca^{2+} and Mg^{2+} in drinking water wells and study their relationship with other hydrogeochemical properties and lithological properties of aquifers using geochemical and stable isotope techniques.

6.2 CARBONATES IN SLOVENIA

Slovenia is diverse country, both geologically and geographically. Elevation varies up to 2,864 m, and the precipitation is relatively high and decreases from west to the east from about 3,000 mm/year to 800 mm/year, with an average of 1,500 mm/year. Consequently, the groundwater resources are rather large, and based on estimates from 2002, only about 15% of available reserves were used (Andjelov *et al.*, 2006).

Karst rocks cover approximately 43% of the total Slovenia surface area, and about 8% belong to dolomites (Figure 6.1). There exist several different limestone and dolomite aquifers among karst rocks and their hydrogeological and geochemical properties depend mostly on their evolution and sedimentological properties (Tucker *et al.*, 1990; Purser *et al.*, 1994). Their hydrogeological role and fractal properties have been described in previous works (Verbovšek, 2008; Verbovšek and Veselič, 2008; Verbovšek, 2009). For detailed sedimentological, diagenetic and hydrogeological properties of dolomites, reader is addressed to refer to these papers and references within, and the major dolomites aquifers are described below (abbreviations are used in the figures and tables):

- Upper Permian dolomite (P_3): early and late-diagenetic relatively pure bedded dolomite, with occasional gypsum lenses.
- Lower Triassic dolomite (T_1): bedded dolomite with high clastic (impurity) content (Dolenec *et al.*, 1981).
- Anisian Dolomite (T_2^1): massive, pure late-diagenetic dolomite.
- Cordevolian dolomite ($_1T_3^1$): massive, late-diagenetic dolomite with a high primary porosity, the greatest grain size and also intensively fractured. Late diagenesis tends to increase the porosity and permeability, due to increase in grain size and consequently intergranular porosity.
- 'Main' Dolomite (T_3^{2+3} M): also known as Dolomia principale in Italy and Hauptdolomit in Germany/Austria – high carbonate and dolomite content, very thick-bedded (Ogorelec and Rothe, 1993; Tucker *et al.*, 1990).

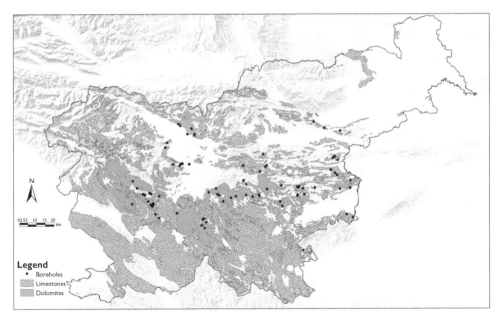

Figure 6.1 Geological map of dolomites and limestones in Slovenia and sampling points (circles). Dolomite and limestone extent is digitized after Basic Geological Map OGK of (ex-)Yugoslavia in scale 1:100,000, topographic map provided by Geodetic Survey of Slovenia (GURS).

- 'Bača' dolomite (T_3^{2+3} B): intensely silicified, late-diagenetic dolomite with lenses and layers of cherts, occuring only in Slovenia's Internal Dinarides tectonic unit (Rožič et al., 2009; Gale, 2010).
- 'Dachstein' limestone/dolomite (T_3^{2+3} D): mostly represented by limestones and dolomitized limestones, lateral equivalent to 'Main' and 'Bača' dolomites, occuring mostly in high Alpine area.
- Other aquifers (Ogorelec et al., 2000), noted in tables and figures, appear with very low number of data, and for this reason are not significant and not discussed further in this paper. Additionally, a multi aquifer mixing may occur ($J_1 + T_3^{2+3}$, $T_1 + M$, $P_2 + P_3$) or unknown classifications (T_2^2, $^1T_3^1$?, T_2, and T_3^{2+3} D).

6.3 APPROACH AND METHOD (SAMPLING PROTOCOLS AND FIELD MEASUREMENTS)

6.3.1 Water sampling

Sampling of groundwater was performed in spring of 2012, in 90 boreholes in mostly dolomite aquifers in the central and southern Slovenia (Figure 6.1). A database of 430 boreholes in dolomite and limestone aquifers has been constructed, out of which 90 boreholes used for public water supply systems, were chosen for hydrogeochemical sampling, as they were more or less continuously pumped to obtain the representative samples. Of these, 78 have been identified with lithological properties (aquifer age and type), and for this reason, 78 of 90 are included in the figures and tables where the results are grouped by lithology. Physicochemical field parameters (temperature, pH, conductivity and Dissolved Oxygen – D.O.) were measured immediately in the field. Sample aliquots collected for cation (pre-treated with HNO_3), anion and alkalinity analyses were passed through a 0.45 µm nylon filter, filled in HDPE bottles and kept refrigerated until analyzed. Samples for $\delta^{13}C_{DIC}$ analyses were stored in glass serum bottles with no headspace and sealed with septa caps.

6.3.2 Laboratory analyses

Alkalinity was measured using Gran titrations (Clesceri et al., 1998). For waters investigated in Sava River watershed it was found out that regression between alkalinity and DIC concentrations was 96% when both were measured (Kanduč et al., 2007). Cations (Ca^{2+}, Mg^{2+}, Na^+, K^+) and anions (SO_4^{2-}, NO_3^-, Cl^-) were measured by ICP-OES (Inductively Coupled Plasma-Optical Emission Spectrometry), IC (Ion Chromatography) and trace elements with ICP-MS (Inductively Coupled Plasma-Mass Spectrometry) in ActLabs Ltd. laboratories, Canada. The stable isotope composition of dissolved inorganic carbon ($\delta^{13}C_{DIC}$) was determined with a Europa Scientific 20–20 continous flow IRMS (isotope ratio mass spectrometer) with an ANCA-TG preparation module. Phosphoric acid (100%) was added (100–200 µl) to a septum-sealed vial which was then purged with pure He. Water sample (6 ml) was injected into the septum tube and headspace CO_2 was measured (modified after Miyajima et al., 1995; Spötl, 2005). A standard solution of Na_2CO_3 (Carlo Erba) with a known $\delta^{13}C_{DIC}$ of −10.8 ± 0.2‰ was used in batch measurement. Program PHREEQC was used to calculate SI (Saturation Indexes) of calcite and dolomite and partial pressure (P_{CO_2}). Program STATISTICA 10.0 was used to evaluate distribution and correlation between measured variables.

6.4   RESULTS

6.4.1   Chemical composition and stable isotope composition of groundwaters

Temperatures in groundwater range from 6.0°C to 15.2°C, conductivity range from 293.8 to 777.2 μS/cm and pH from 7.35 to 8.22. Major geochemistry data are presented in Table 6.1 according to statistics.

Total alkalinity ranged from 2.1 to 8.6 mM, Ca^{2+} range from 0.57 to 2.57 mM and Mg^{2+} from 0.27 mM to 1.83 mM. Groundwater chemistry is compared to Sava River in Slovenia (Kanduč *et al.*, 2007). Groundwater in our study is saturated with oxygen, therefore redox reactions e.g. denitrification, sulphate reduction, and methanogenesis are not likely to happen in groundwater.

Calculated partial pressures (P_{CO_2}) in groundwater varied from 1000 ppm to 17790 ppm. The calcite saturation index ($SI_{calcite} = \log ([Ca^{2+}]\cdot[CO_3^{2-}]/K_{calcite}$; where $K_{calcite}$ is solubility product of calcite)) and dolomite ($SI_{dolomite} = \log ([Mg^{2+}]\cdot[Ca^{2+}]\cdot[CO_3^{2-}]/K_{dolomite}$; where $K_{dolomite}$ is solubility product of dolomite)) saturation index ranged from −0.04 to 0.98 for calcite and from −0.21 to 1.76 for dolomite (Tables 6.1), respectively. Practically all waters are therefore slightly oversaturated with both calcite and dolomite.

Further, chemical composition of groundwater in relation to age of carbonate rocks was evaluated with statistical methods. Isotopic composition of inorganic carbon ($\delta^{13}C_{DIC}$) range from −14.6 to −8.2‰ and depends on soil CO_2 contribution in groundwater.

6.4.2   Statistical distribution and values

Distribution of Ca^{2+} and Mg^{2+} can be attributed either to normal or lognormal distribution (Figure 6.2A and 6.2B), contrary to most other elements, which follow lognormal distribution. Magnesium tends to be more normally distributed, and calcium more lognormally.

Number of data varies greatly among the different aquifers (Table 6.1), and is the greatest in Cordevolian dolomites, 'Main' and 'Bača' dolomites and Anisian dolomites. An outlier can be identified in the borehole, drilled in Lower Triassic rocks, as it has anomalously high values of Ca^{2+} (313 mg/L) and Mg^{2+} (72 mg/L) with pH equal to 7.82, indicating dissolution of gypsum/anhydrite. This is confirmed by very high concentration of SO_4^{2-} (943 mg/L compared to median value of 8.5 mg/L in all boreholes). With this exception, all water samples have concentrations lower than permissible limit for drinking water according to Slovenian legislation (Rules on Drinking Water, Official Gazette of Republic of Slovenia, no. 19/2004).

Several differences among aquifers exist (Figure 6.3), and can be attributed to geological properties of aquifers. Considering the molar ratio Mg^{2+}/Ca^{2+}, apart from

Table 6.1 Mean values of measured ions, pH, (TDS), Saturation indexes (SI) of calcite and dolomite and partial pressure of CO_2. All concentration units are in mg/L, except pH and SI (without units) and PCO_2 (bar). N: number of data.

Geology	N	TDS	pH	HCO_3^-	Mg^{2+}	Ca^{2+}	Na^+	K^+	Cl^-	SO_4^{2-}	Si	SI_{cal}	SI_{dol}	pCO_2
T_3^{2+3} M	33	353.65	7.78	390.66	32.27	61.21	1.818	0.525	40.74	7.58	1.610	0.53	0.91	-2.21
T_3^1	12	380.67	7.80	381.68	34.38	65.32	1.309	0.507	2.17	12.73	2.058	0.56	0.98	-2.25
T_3^2	11	317.17	7.82	323.79	27.93	57.93	2.517	0.819	3.50	12.72	2.245	0.49	0.82	-2.33
T_2^2	1		7.83	356.97	22.00	64.90	1.770	0.490	1.10	15.60	2.600	0.62	0.93	-2.28
$J_1 + T_3^{2+3}$	1	232.90	8.16	261.78	19.60	39.90	1.450	0.580	3.03	1.69	1.400	0.57	0.94	-2.76
T_1	2	776.00	7.98	265.13	46.90	183.80	6.330	1.200	2.89	479.10	6.750	0.86	1.36	-2.60
$T_1 + M$	2	507.35	7.38	411.27	27.40	98.35	19.350	3.085	31.70	28.95	5.100	0.42	0.48	-1.77
T_3^{2+3} B	6	383.77	7.84	381.88	32.88	71.43	2.883	0.878	3.09	12.07	3.517	0.66	1.15	-2.27
T_3^1?	1	466.60	7.62	483.28	44.50	80.60	1.330	0.850	1.89	23.00	2.900	0.59	1.07	-1.95
T_2	4	370.88	7.72	360.63	32.15	57.35	1.984	5.010	4.10	7.36	1.700	0.43	0.75	-2.19
$P_2 + P_3$	1	263.60	7.93	136.07	21.80	40.50	1.440	0.420	2.10	19.20	3.000	0.13	0.14	-2.80
P_3	1	308.10	7.91	209.91	35.90	32.20	1.810	0.300	6.20	35.20	5.500	0.16	0.49	-2.60
T_3^{2+3} D?	3	217.30	8.08	203.81	11.97	46.87	1.853	0.403	4.51	4.34	1.300	0.43	0.32	-2.83
All Grps	**78**	**365.29**	**7.80**	**361.28**	**31.27**	**64.90**	**2.479**	**0.899**	**19.76**	**22.67**	**2.239**	**0.53**	**0.88**	**-2.28**

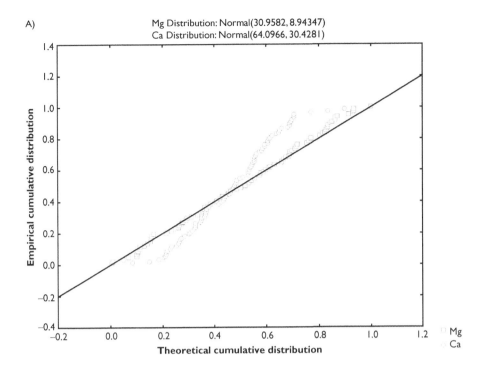

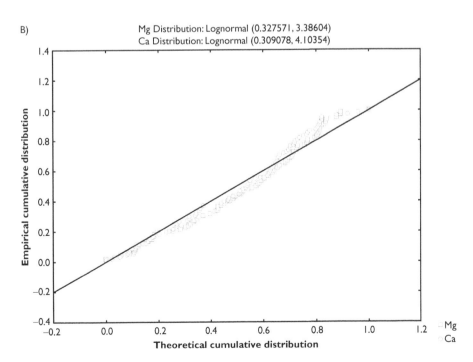

Figure 6.2 Distribution of Ca²⁺ and Mg²⁺ ions in water, in probability-probability Q-Q plots. A) fitting to normal distribution, B) fitting to lognormal distribution.

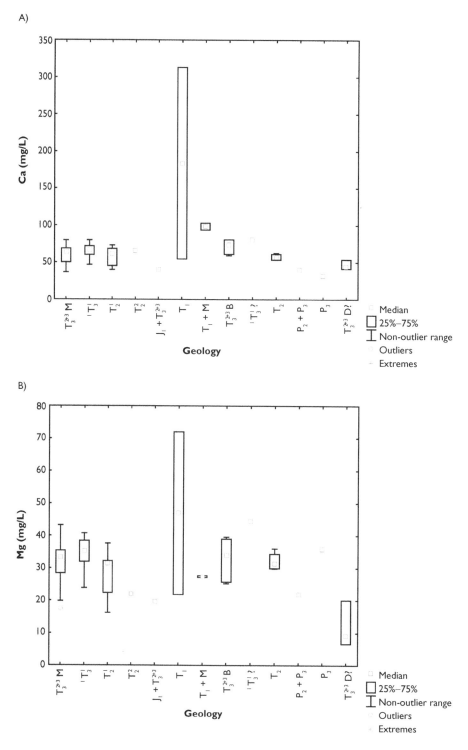

Figure 6.3 (Continued).

C)

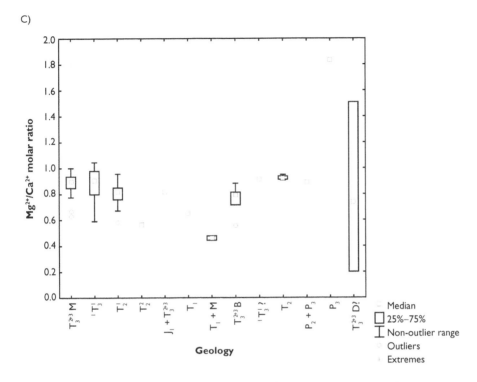

Figure 6.3 Box-plot diagram of A) Ca^{2+} and B) Mg^{2+} concentrations in samples water.

Dachstein limestones/dolomitized limestones with understandable greatest variation of Mg^{2+}/Ca^{2+} ratios. Cordevolian ($_1T_3^1$) dolomites are the most 'pure' dolomites, being closest to molar ratio of 1.00. Similar ratios are found in 'Main' (T_3^{2+3} M.) and Anisian (T_2^1) dolomites; both being early- and late-diagenetic dolomites of very high carbonate content (over 95%).

6.5 DISCUSSION

6.5.1 Major geochemistry of karst fissured aquifers

Figure 6.4 presents $Ca^{2+} + Mg^{2+}$ versus alkalinity. All samples from our study are close to line 2:1 mole ratio of HCO_3^- to $Ca^{2+} + Mg^{2+}$, meaning that cations are leached from carbonates (Eqs. 6.1 and 6.2). $Ca^{2+} + Mg^{2+}$ ratio in our study ranges from 1.43 to 3.83 mM, which is in wider range in comparison to Sava River ($Ca^{2+} + Mg^{2+}$ range from 0.6 to 1.6 mM) and is reasonable since groundwaters are much more mineralized due to longer water–rock interactions during groundwater flow in comparison to surface waters. Deviations from carbonate dissolution line 2:1 (Figure 6.4) are probably due to weathering of other minerals such as albite and anorthite, which are weathered to clay minerals (unsoluble fraction of carbonates) and also contribute to alkalinity.

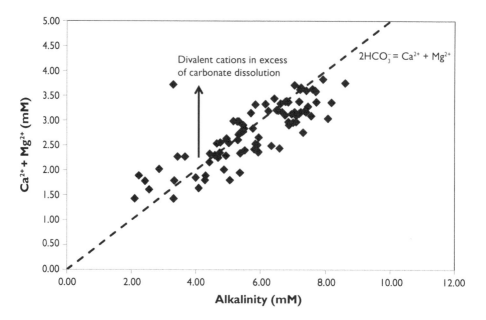

Figure 6.4 Ca²⁺ + Mg²⁺ ratio versus alkalinity with lines 1:2 indicating weathering of carbonates within aquifers.

Figure 6.5 shows Mg^{2+} versus Ca^{2+} values to determine the relative contribution of dolomite versus calcite to carbonate weathering intensity in groundwater. It can be observed that dolomite dominates the water chemistry of groundwaters. It was previously found that Slovenian streams have a wide range of Mg^{2+}/Ca^{2+} ratios, indicating variations in the relative contributions of calcite and dolomite in different tributaries, ranging from 0.2 to 0.8 (Szramek *et al.*, 2007). The Mg^{2+}/Ca^{2+} ratio observed in Alpine springs is in the range of 0.07 to 0.37, indicating lower Mg^{2+} contribution to headwaters recharging the Sava River Basin (Kanduč *et al.*, 2012).

6.5.2 Evaluation of geochemical processes and contribution of dissolution of carbonates versus degradation of organic matter in karst-fissured aquifers

$\delta^{13}C_{DIC}$ (isotopic composition of inorganic carbon) values range from −14.6 to −8.2‰ and reveal the contribution of organic matter and dissolution of carbonates within aquifer. Lower $\delta^{13}C_{DIC}$ values in aquifers reveal the presence of thicker soils (more soil CO_2 contribution) in recharge areas. Therefore, the most vulnerable aquifers are ones with low soil CO_2 contributions (Kanduč *et al.*, 2012). The highest $\delta^{13}C_{DIC}$ value is observed in aquifer Zavčen, menaning that dissolution of carbonates dominate degradation of organic matter, while the lowest $\delta^{13}C_{DIC}$ is observed within aquifer ČG-1/02

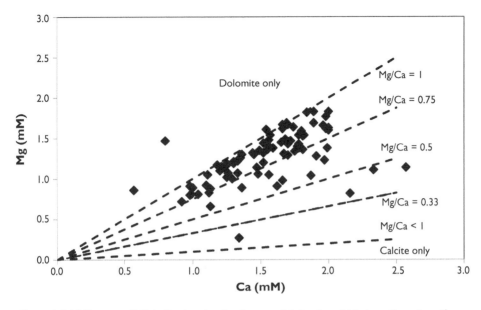

Figure 6.5 Mg²⁺ versus Ca²⁺ indicating the dominance of dolomite within investigated aquifers.

indicating that degradation of organic matter is the dominant process within aquifer (Figure 6.6). An average $\delta^{13}C_{POC}$ value of $-27.0‰$ was assumed to calculate lines 1–3 in Figure 6.6. Open system equilibration of DIC with soil CO_2 enriches DIC in ^{13}C by about 9‰ (Mook *et al.*, 1974), thus yielding the estimate of $-18.0‰$ shown on Figure 6.6. Nonequilibrium dissolution of carbonates with one part of DIC originating from soil CO_2 with $-27.0‰$ and other from carbonate dissolution with an average $\delta^{13}C_{ca}$ of 3.2‰ produces an intermediate $\delta^{13}C_{DIC}$ value of $-11.9‰$ (Figure 6.6). Considering average isotopic composition of carbonates ($\delta^{13}C_{ca}$) comprising in the recharge area and isotopic fractionation due to dissolution of carbonates, which is $1.0‰ \pm 0.2‰$ enrichment with ^{12}C (Romanek *et al.*, 1992) the $\delta^{13}C_{DIC}$ value would be $2.2‰ \pm 0.2‰$ (line 3, Figure 6.6).

$\delta^{13}C_{DIC}$ value can determine contributions of organic matter decomposition and carbonate mineral dissolution within aquifer. $\delta^{13}C_{DIC}$ in aquifer is also controlled by the geological composition of the recharge and depth of soil through which water is infiltrating. Higher soil thickness means more CO_2 contribution from organic matter meaning lower $\delta^{13}C_{DIC}$ values measured within aquifer (Kanduč *et al.*, 2012).

A simple mass balance calculation in groundwaters is presented (Eqs. 6.3 and 6.4), considering the following processes: organic matter degradation (DIC_{org}) and carbonate dissolution (DIC_{ca}) as groundwaters are considered as systems without atmospheric exchange (closed system):

$$1 = DIC_{org} + DIC_{ca} \tag{6.3}$$
$$DIC_{groundwater} \cdot \delta^{13}C_{DIC} = DIC_{org} \cdot \delta^{13}C_{org} + DIC_{ca} \cdot \delta^{13}C_{ca} \tag{6.4}$$

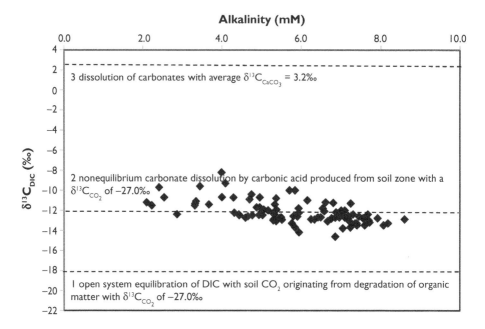

Figure 6.6 $\delta^{13}C_{DIC}$ values with lines indicating geochemical processes within aquifer. These include: 1. dissolution of carbonates according to the average $\delta^{13}C_{CaCO_3}$ (3.2‰) (Kanduč *et al.*, 2007); 2. nonequilibrium carbonate dissolution by carbonic acid produced from soil zone CO_2; and 3. open system equilibration of DIC with soil CO_2 originating from degradation of organic matter with $\delta^{13}C_{soil} = -27.0‰$, which is characteristic for Slovenian soils (Kanduč *et al.*, 2007).

$\delta^{13}C_{POC}$ and $\delta^{13}C_{Ca}$ measured values of $-27‰$ and $3.2‰$ (Kanduč *et al.*, 2007, 2012) were taken into account in the mass balance equations. The contribution of rainwater in mass balance is considered to be minimal as it contains only a small amount of DIC (Yang *et al.*, 1996). According to mass balance calculation, it was estimated that contribution of organic matter within aquifer ranged from 41.0 to 62.2% and dissolution of carbonates from 37.8 to 59‰.

6.6   CONCLUSIONS

Presented groundwater solute chemistry is dominated by HCO_3^-, Ca^{2+} and Mg^{2+}. Individual aquifers are varied due to their geochemical and stable isotope composition. Total alkalinity ranges from 2.1 to 8.6 mM, Ca^{2+} from 0.57 to 2.57 mM and Mg^{2+} from 0.27 to 1.83 mM. Investigated groundwater in our study was oversaturated with pCO_2 (representing a source of CO_2 to atmosphere) and calcite and dolomite. Weathering of dolomite controls the major ion geochemistry of groundwater. The biogeochemical processes affecting DIC concentrations and $\delta^{13}C_{DIC}$ values were quantified by mass balance calculation, showing that the most important processes in all investigated groundwaters are degradation of organic matter and dissolution of carbonates. Vulnerability of groundwater is highly related with soil profiles, meaning thicker soils lead to longer infiltration times, which reduces vulnerability to surface contaminants.

Stable isotopes are useful tool in solving problems in groundwater protection zones to prevent future pollution and estimate aquifer vulnerabilities for risk assessment studies. Therefore geochemical and stable isotope techniques represent complementary approach to the EU Water Framework Directive.

ACKNOWLEDGEMENTS

This project has been funded by Slovenian Research Agency (ARRS), project number Z1-3670, and programme research group P1-0143.

REFERENCES

Andjelov M., Gale U., Kukar N., Trišič N. & Uhan J. (2006) Ocena količinskega stanja podzemnih voda v Sloveniji (Groundwater quantitative status assessment in Slovenia). *Geologija* 49/2:383–391.

Atekwana E.A. & Krishnamurthy R.V. (1998) Seasonal variations of dissolved inorganic carbon and $\delta^{13}C$ of surface waters: application of a modified gas evaluation technique. *Journal of Hydrology* 205(3–4):260–278.

Aucour A.M., Sheppard S.M.F., Guyomar O.J. & Wattelet J. (1999) Use of ^{13}C to trace origin and cycling of inorganic carbon in the Rhône river system. *Chemical Geology* 159(1–4):87–105.

Clesceri L.S., Greenberg A.E. & Eaton A.D. (1998) *Standard methods for the examination of water and wastewater.* 20th edition. APHA, AWWWA, WEF, Baltimore, 1325p.

de Marsily G., Delay F., Gonçalvès J., Renard Ph., Teles V. & Violette S. (2005) Dealing with spatial heterogeneity. *Hydrogeology Journal* 13(1):161–183.

Dolenec T., Ogorelec B. & Pezdič J. (1981) Zgornjepermske in skitske plasti pri Tržiču (Upper Permian and Scythian beds in the Tržič area). *Geologija* 24(2):217–238.

Drever J.I. (1997) *The geochemistry of natural waters: surface and groundwater environments.* Prentice-Hall, New Jersey, 436p.

Gale, L. (2010) Microfacies analysis of the Upper Triassic (Norian) "Bača Dolomite": early evolution of the western Slovenian Basin (eastern Southern Alps, western Slovenia). *Geologica Carpatica* 61(4):293–308.

Kanduč T., Szramek K., Ogrinc N. & Walter L.M. (2007) Origin and cycling of riverine inorganic carbon in the Sava River watershed (Slovenia) inferred from major solutes and stable carbon isotopes. *Biogeochemistry* 86:137–154.

Kanduč T., Mori N., Kocman D., Stibilj V. & Grassa F. (2012) Hydrogeochemistry of Alpine springs from North Slovenia: insights from stable isotopes. *Chemical Geology* 300/301:40–54.

Langmuir D. (1997) *Aqueous environmental geochemistry.* Prentice-Hall, New Jersey, 600p.

Li S.L., Liu C.Q., Tao F.X., Lang Y.C. & Han G.L. (2005) Carbon Biogeochemistry of Ground Water, Guiyang, Southwest China. *Ground Water* 43:494–499.

Miyajima T., Yamada Y. & Hanba Y.T. (1995) Determining the stable isotope ratio of total dissolved inorganic carbon in lake water by GC/C/IRMS. *Limnology and Oceanography* 40(5):994–1000.

Mook W.G., Bommerson J.C. & Staverman W.H. (1974) Carbon isotope fractionation between dissolved bicarbonate and gaseous carbon dioxide. *Earth and Planetary Science Letters* 22:169–176.

Official Gazette of Republic of Slovenia (Uradni list) (2004) Rules on Drinking Water. http://www.uradni-list.si/1/objava.jsp?urlid=200419&stevilka=865.

Ogorelec B. & Rothe P. (1993) Mikrofacies, Diagenese und Geochemie des Dachsteinkalkes und Hauptdolomits in Süd-West-Slowenien (Microfacies, Diagenesis and Geochemistry of Dachstein limestone and Main Dolomite in Southwestern Slovenia). *Geologija* 35:81–181.

Ogorelec B., Dolenec T. & Pezdič J. (2000) Izotopska sestava O in C v mezozojskih karbonatnih kamninah Slovenije (Isotope composition of O and C in Mesozoic carbonate rocks of Slovenia). *Geologija* 42:171–205.

Purser B.H., Brown A. & Aissaoui D.M. (1994) Nature, origins and evolution of porosity in dolomites. *In:* Purser B., Tucker M. & Zenger D. (eds) *Dolomites*. The International Association of Sedimentologists Special Publication 21. The International Association of Sedimentologists, Cambridge, 283–308.

Romanek C.S., Grossman E.L. & Morse J.W. (1992) Carbon isotopic fractionation in synthetic aragonite and calcite: effects temperature and precipitation rate. *Geochimica et Cosmochimica Acta* 46:419–430.

Rožič, B., Kolar-Jurkovšek, T. & Šmuc, A. (2009) Late Triassic sedimentary evolution of Slovenian Basin (eastern Southern Alps): description and correlation of the Slatnik Formation. *Facies* 55(1):137–155, doi: 10.1007/s10347-008-0164-2.

Spötl C. (2005) A robust and fast method of sampling and analysis of $\delta^{13}C$ of dissolved inorganic carbon in ground waters. *Isotopes in Environmental and Health Studies* 41:217–221.

Szramek K., Walter L.M., Kanduč T. & Ogrinc N. (2011) Dolomite versus calcite weathering in hydrogeochemically diverse watersheds established on bedded carbonates (Sava and Soča Rivers, Slovenia). *Aquatic Geochemistry* 17:357–396.

Szramek K., McIntosh J., Williams E., Kanduč T., Ogrinc N. & Walter L.M. (2007) Relative weathering intensity of calcite versus dolomite in carbonate-bearing temperature zone watersheds: carbonate geochemistry and fluxes from catchments within the St. Lawrence and Danube river basins. *Geochemistry, Geophysics, Geosystems* 26(4):1–26.

Tucker M.E., Wright V.P. & Dickson J.A.D. (1990) *Carbonate Sedimentology*. Blackwell Scientific Publications, Oxford, 496p.

Verbovšek T. (2008). Diagenetic effects on well yield of dolomite aquifers in Slovenia. *Environmental Geology* 53(6):1173–1182, doi:10.1007/s00254-007-0707-9.

Verbovšek T. (2009) Influences of aquifer properties on flow dimensions in dolomites. *Ground Water* 47(5):660–668.

Verbovšek T. & Veselič M. (2008) Factors influencing the hydraulic properties of wells in dolomite aquifers in Slovenia. *Hydrogeology Journal* 16(4):779–795.

Williams E.L., Szramek K.J. & Jin L. (2007) The carbonate system geochemistry of shallow groundwater-surface water systems in temperate glaciated watersheds (Michigan, USA): significance of open-system dolomite weathering. *Geological Society of America Bulletin* 119:515–528.

Yang C., Telmer K. & Veizer J. (1996). Chemical dynamics of the 'St. Lawrance' riverine system: δD_{H2O}, $\delta^{18}O_{H2O}$, $\delta^{13}C_{DIC}$, $\delta^{34}S_{sulfate}$, and dissolved $^{87}Sr/^{86}Sr$. *Geochimica et Cosmochimica Acta* 60:851–866.

Zavadlav S., Kanduč T., McIntosh J. & Lojen S. (2013) Isotopic and Chemical Constraints on the Biogeochemistry of dissolved Inorganic Carbon and Chemical weathering in the Karst Watershed of Krka River (Slovenia). *Aquatic Geochemistry* 19(3):209–230. doi: 10.1007/s10498-013-9188-5.

Chapter 7

Distribution of Ca and Mg in groundwater flow systems in carbonate aquifers in Southern Latium Region (Italy): Implications on drinking water quality

Giueseppe Sappa[1], *Sibel Ergul*[1] *& Flavia Ferranti*[2]

[1]*Dipartimento di Ingegneria Civile, Edile ed Ambientale,*
Sapienza – Università di Roma, Roma, Italy
[2]*Centro Reatino di ricerche di Ingegneria per la Tutela e la*
Valorizzazione dell'Ambiente e del Territorio,
Sapienza – Università di Roma, Rieti, Italy

ABSTRACT

The investigation of hydrochemical properties of springs from the carbonate aquifers of the southern Latium region (Central Italy) aimed to: (i) identify different processes responsible for the hydrochemical evolution of groundwater; (ii) determine the levels of Ca and Mg in the drinking water networks and; (iii) determine the effect of hardness on water quality. Based on the dominance of major cations and anions three hydrochemical facies have been identified: (1) Ca–Mg–HCO_3; (2) Mixed Ca–Na–HCO_3–Cl; (3) Na–Cl. In all cases, Ca–Mg–HCO_3 facies predominates reflecting the main rock types in the area, where limestone, dolomitic limestones and dolomites are common. The Total Dissolved Solids (TDS) and Electrical Conductivity (EC) range from 101 to 1320 mg/l and 138 to 2310 µS/cm, respectively. Most of the springs are supersaturated with respect to carbonate minerals, however all sampled waters were undersaturated with respect to evaporite minerals. The compositional changes in Ca and Mg concentrations are controlled by distance from the recharge area (i.e. groundwater discharging from the lower elevations tends to have the highest concentrations of Mg). The classification of water based on total hardness (as $CaCO_3$) shows that most of the spring water samples fall between hard (150–300 mg/l) and very hard (>300 mg/l) water type.

7.1 INTRODUCTION

As one of the most important water supply sources worldwide, spring water and groundwater from carbonate aquifers have significant importance. Many studies have been carried out on carbonate aquifer systems including investigations of the geochemical processes and their hydrogeological implications. The chemical composition of groundwater is controlled by many factors, including the composition of the precipitation, geological structure and the mineralogy of the aquifers (Chenini & Khmiri, 2009). The interaction of all these factors generates a variety of water types. In recent years, geochemical modelling methods have been employed to obtain information from hydrochemical datasets in the aquifer systems. These techniques can help to resolve hydrogeological factors such as the hydrochemical components along the groundwater flow paths and geochemical controls on water quality (Briz Kishore & Murali, 1992).

The chemistry of water from carbonate aquifers and the variations in hydrochemical properties are generally related to (i) water-rock interactions (ii) natural factors such as mixing between seawater and freshwater (iii) anthropogenic factors (iv) the type of groundwater circulation. The composition of water in carbonate systems is the result of the dissolution of variable quantities of carbonate (i.e. calcite and dolomite) and evaporite minerals (i.e. gypsum and halite) that control the water chemistry (White, 1988; Ettazarini, 2005). The high dissolution rate and residence time of the circulating water in carbonate rocks and ongoing dissolution allows the water to become more saturated with respect to carbonate minerals along the groundwater flow paths (Plummer, 1977). Consequently, groundwater composition that is controlled by carbonate reactions has a relatively high Ca and HCO_3 concentration, and, if the rock includes some dolomite, could also have quite high Mg concentrations (Shuster & White, 1971). Ca and Mg are the most important elements essential for human health. The presence of Ca and Mg in drinking water has important beneficial effects on human health, however at very high levels they can cause some negative aesthetic effects (undesirable taste and colour). Moreover, water with over 125 mg-Mg/l can cause a laxative effect (Johnson & Scherer, 2012). These elements are the principal natural sources of hardness in water. Many studies have been carried out to emphasise the importance of Ca, Mg and concentrations of hardness in drinking water in relation to human health. These studies suggest that Ca and Mg in drinking water are important protective factors against colon cancer, cardiovascular and cerebrovascular disease and acute myocardial infarction (Ferrandiz et al., 2004; Rubenowitz et al., 1996; Yang & Hung, 1998).

Groundwater resources in the southern part of the Latium region play an important role in providing water for domestic, industrial and agricultural uses. Increased knowledge of geochemical processes that control groundwater chemistry from the aquifer systems can contribute to improved understanding of the processes influencing the composition of water within the carbonate aquifers. The main objectives of this chapter are (i) to identify the source and distribution of calcium and magnesium in water and (ii) its effect on drinking water quality comparing the parameters (i.e. total hardness, TDS, EC, Ph) with the standard guideline values.

7.2 METHODOLOGY

The main spring water sampling survey was carried out in the Lepini, Ausoni and Aurunci Mountains, between 2002 and 2004. On the basis of the hydrogeological setting of the area, 54 spring samples were characterised. Water temperature, electrical conductivity and pH values were determined in the field using a PC 300 Waterproof Hand-held meter. Bicarbonate was measured by titration with 0.1 N HCl. Water samples were filtered through cellulose filters (0.45 μm), and their major and minor constituents were determined by a Metrohm 761 Compact IC ion chromatograph (replicability ±2%). A Metropes C2-100 column was used to determine cations (Na^+, K^+, Mg^{++}, Ca^{++}), while a Metropes A Supp 4-250 column was used for anions (Cl^-, SO_4^-, HCO_3^-). Chemical analyses were performed on the water samples at the

Geochemical Laboratory of Sapienza, University of Rome. The analytical accuracy of these methods ranged from 2 to 5%. The geochemical program PHREEQC software, version 2.10.0.0 (Parkhurst & Appello, 1999), with the thermodynamic dataset wateq4f.dat, was employed to evaluate the saturation status of minerals in spring water samples. The SI (Saturation Index) indicates the potential for chemical equilibrium between water and minerals and the tendency for water-rock interaction (Wen, 2008). The characterisation of spring and well water samples has been evaluated by means of major ions, Ca^{++}, Mg^{++}, HCO_3^-, Na^{++}, K^+, Cl^-, SO_4^-, as they are the best indicators of chemical evolution along groundwater flow paths. For the identification of water types, the chemical analysis data of the spring water samples have been plotted on a Piper diagram using Geochemistry Software AqQA.

7.3 STUDY AREA

Lepini, Ausoni and Aurunci are three different groups of mountains belonging to the pre-Apennines of Latium and they occupy a well-defined geographical area, called the Volscian Mountain Range (Figure 7.1). The Lepini Mountains are located in the northern part of the Pontina Plain and host an important karst aquifer. The aquifer in the Lepini massif is unconfined with an undefined depth. The Pontina Plain is a

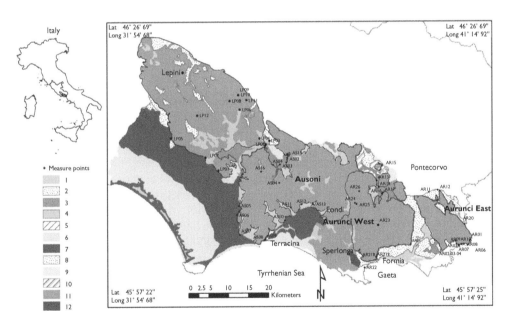

Figure 7.1 Simplified hydro-geological map of the study area. 1. Recent deposits (Holocene); 2. Detritic complex (Pleistocene-Holocene); 3. Alluvial complex (Pleistocene-Holocene); 4. Alluvial deposits from perennial streams (Pleistocene-Holocene); 5. Travertine complex (Pleistocene-Holocene); 6. Sand dunes (Pleistocene-Holocene); 7. Fluvial lacustrine deposits (Holocene); 8. Pyroclastic complex (Pliocene-Pleistocene); 9. Heterogeneous clastic deposits (Pleistocene); 10. Clayey-marly Flysch complex with interbedded lithoids (Cretaceous – Miocene); 11. Carbonate platform complexes (Middle Lias-Upper Cretaceous); 12. Basal dolomite complex (Triassic-lower Lias). Spring and well sampling locations (LP: Lepini springs., AS: Ausoni springs., AR: Aurunci springs).

Table 7.1 Summary statistics of major ion concentrations and physico-chemical parameters of sampled waters.

Sampling locations		$T°C$	pH	EC ($\mu s/cm$)	Ca (mg/l)	Mg (mg/l)	Na (mg/l)	K (mg/l)	Cl (mg/l)	HCO_3^- (mg/l)	SO_4^{-2} (mg/l)	TDS (mg/l)
Lepini springs	Mean	13.0	7.7	517.0	64.3	13.7	37.4	2.9	55.4	239.8	16.4	430
	Median	13.0	7.7	400.0	67.3	6.6	6.8	1.2	9.6	235.9	4.3	334
	Min	10.0	6.9	138.0	15.4	1.4	2.9	0.1	3.9	67.1	1.7	101
	Max	15.0	8.1	1540.0	111.0	44.7	221.0	15.8	338.4	448.0	85.4	1264
Ausoni springs	Mean	13.0	7.7	826.0	65.2	18.2	73.5	3.5	128.4	234.2	27.2	550
	Median	12.0	7.7	404.0	61.6	9.2	8.6	0.8	13.3	232.0	5.8	324
	Min	12.0	6.9	315.0	41.5	3.8	4.1	0.2	7.5	176.9	3.8	255
	Max	15.0	8.1	2310.0	89.2	47.8	293.1	15.4	524.9	305.1	110.9	1320
Aurunci springs	Mean	12.0	7.7	545.0	70.7	25.6	10.6	2.9	12.7	316.5	34.6	479
	Median	12.0	7.8	429.0	63.6	9.9	7.6	1.1	9.8	244.1	5.3	338
	Min	3.0	7.1	311.0	44.5	1.5	4.2	0.3	4.5	170.9	2.7	246
	Max	31.0	8.0	1217.0	197.3	93.4	50.5	21.6	46.7	805.5	195.8	1150

coastal plain developed along an extensional marine boundary. This Plain is positioned between the Lepini-Ausoni mountains of the Central Appenines and the Tyrrhenian Sea. The Ausoni Mountains rise in southern Latium and extend towards the coast, from the middle Amaseno valley. The Ausoni hydrogeological unit is mainly composed of limestones with interbedded dolomitic limestones. Most of the springs lie along the periphery of the aquifer with no sharp distinction from the recharge areas. The Aurunci Mountains represent the south eastern part of the Volscian Range and are oriented parallel to the Apennine Range. The Aurunci Mountains are made of two distinct hydrogeological units: the western Aurunci, belonging to the Ausoni-Aurunci system, and the eastern Aurunci, which is separated from the western part by a marly-arenaceous flysch complex (Boni, 1975). The Western Aurunci hydrogeological unit consists of dolomitic limestones and dolomites of Jurassic and Cretaceous age. The springs are supplied by groundwater derived from these geological formations. The groundwater is directly discharged into the Liri river through the narrow alluvial belt separating the unit from the river. The unit ontains multiple hydrogeological basins and the boundaries match important tectonic lines that caused the outcropping of the calcareous-dolomitic Jura (Accordi *et al.*, 1976). The Eastern Aurunci hydrogeological carbonate structure is surrounded by relatively less-permeable sediments, including the Frosinone flysch, the Roccamonfina volcanites and the Garigliano plain alluvia (Celico, 1978).

7.4 RESULTS & DISCUSSION

7.4.1 Water chemistry and hydrochemical evolution

The summary statistics of major ion concentrations and physico-chemical characteristics of the spring water samples are presented in Table 7.1. The sampled waters in the study area show different characteristics in terms of physico-chemical parameters and elemental concentrations. The temperature of the sampled spring waters range from 3 to 31°C with the mean value of 13.4°C. The highest temperatures (31°C) were observed in two samples from Aurunci springs. The pH values of the water samples range from 6.91 to 8.12, indicating slightly acidic to slightly alkaline nature. According to the WHO (2004) guidelines, the range of desirable pH values for drinking water is 6.5–9.2. There are no spring water samples with pH values outside of the desirable ranges. The Electrical Conductivity (EC) and Total Dissolved Solids (TDS) values range from 315 to 2310 µS/cm and 255 to 1320 mg/l, respectively. The large variation in Total Dissolved Solids (TDS) is thought to be mainly due to water-rock interaction along the flow paths and proximity of the sampling locations to the coast. Most of the spring samples (~61%) show TDS values below 1000 mg/l and EC values less than the maximum permissible drinking water guideline limit highlighting suitability for drinking and agricultural purposes. The highest calcium (197.3 mg/l) and magnesium (93.4 mg/l) concentrations were observed in water samples from Aurunci springs. Bicarbonate is the dominant anion found in all spring water samples, varying from 67.1 mg/l to 805 mg/l. The highest sodium and chloride concentrations were observed in Ausoni and Lepini springs (i.e. samples taken from low discharge elevations) ranging from 293.1 mg/l to 524.9 mg/l and 221 mg/l to 338 mg/l, respectively.

The hydrochemical facies of the spring waters are illustrated in the Piper trilinear diagram (Figure 7.2). Based on the dominance of major cationic and anionic species,

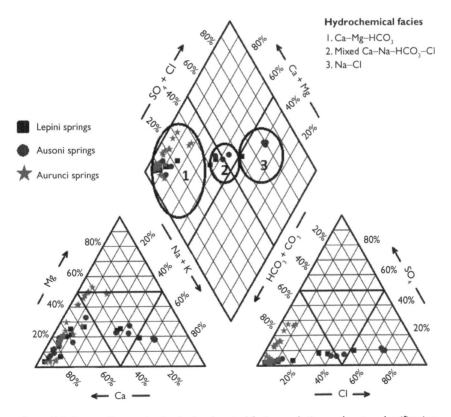

Figure 7.2 Piper trilinear plot for hydrochemical facies evolution and water classification.

three hydrochemical facies have been identified in the study area: (1) Ca–Mg–HCO₃; (2) Mixed Ca–Na–HCO₃–Cl; (3) Na–Cl. The Ca–Mg–HCO₃ water type predominates, reflecting the main rock types in the area, where limestone, dolomitic limestones and dolomites are the most common formations. However, some spring samples discharge at lower elevations, issuing from Lepini and Ausoni Mountains, and belong to or show a tendency towards the Na–Cl type due to the proximity of the sampling locations to the coast. The variations in ion concentrations and the occurrence of different hydrochemical facies can be attributed to water-rock interaction and/or seawater intrusion in the coastal area.

This was confirmed by geochemical modelling and determination of SI for the Lepini, Ausoni and Aurunci springs. The SI indicates the potential for chemical equilibrium between water and minerals and the tendency for water-rock interaction (Jawad *et al.*, 1986). If undersaturated (SI < 0), this phase could be dissolved by the groundwater and, thus, could be a potential source of its constituent chemistry. If supersaturated (SI > 0), that phase, feasibly, could precipitate. Geochemical modelling and saturation index analysis of the Lepini, Ausoni and Aurunci spring samples shows an interaction with carbonate rocks. Calculated saturation indexes with respect to various mineral phases (i.e. calcite, dolomite, gypsum and halite) of the spring water samples are shown in Figure 7.3.

The results of geochemical modelling suggest that most of the spring water samples from the Lepini, Ausoni, and Aurunci Mountains are saturated with respect to

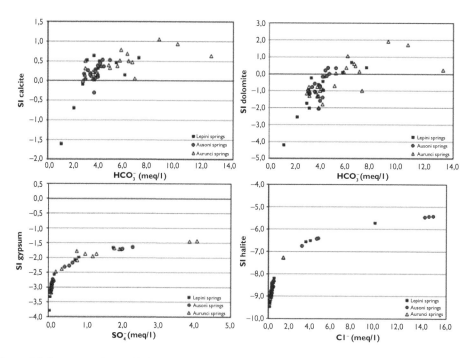

Figure 7.3 Saturation Index (S.I.) values of spring water samples. a) Calcite saturation index versus HCO₃⁻; b) Dolomite saturation index versus HCO₃⁻; c) Gypsum saturation index versus SO₄⁻; d) Halite saturation index versus Cl⁻.

calcite. More than half of the samples are undersaturated with respect to dolomite, while all the sampled waters are undersaturated with respect to gypsum and halite. This indicates that the groundwater has capacity to dissolve gypsum and halite along the flow paths so that the concentrations of Ca^{++}, SO_4^{-}, Na^+ and Cl^- in solution could increase down gradient (Stumm & Morgan, 1996). The spring samples that are both saturated with respect to calcite and dolomite imply a greater dissolution and stronger mineralisation down the groundwater flow paths. However, the water samples, undersaturated with respect to dolomite, indicate that dolomite can dissolve in this system adding Ca^{++}, Mg^{++}, and HCO_3^{-} in solution. If Ca and Mg were derived from the dissolution of carbonate (calcite and dolomite) and evaporate (gypsum) minerals the ionic ratios of (Ca + Mg) to (SO_4 + HCO_3) should be a constant value of one (McLean et al., 2000). Thus, binary plots of (Ca + Mg) versus (HCO_3 + SO_4) were prepared to identify the ion ex-change and weathering processes (Figure 7.4). Most of the spring water samples fall along the 1:1 relationship suggesting that these ions have originated from the dissolutions of calcite, dolomite and gypsum. However, spring samples from the Ausoni Mountains are clustered above the 1:1 line indicating ion exchange process. Na:Cl should be a constant value of one, if halite dissolution is responsible for the sodium and chloride (Fisher & Mullican, 1997). The Na:Cl ratios of spring water samples range from 0.08 to 2.25 (Table 7.2). The majority of the samples have a molar ratio greater or equal to one, which indicates Na^{++} may increase due to ion exchange and halite dissolution (Cerling et al., 1989). However, some spring water samples from the Ausoni Mountains have a Na:Cl molar ratio close to and less than the value in seawater (0.86) indicating possible seawater contamination (where groundwater becomes enriched in Cl^- the value of the ratio will drop).

The compositional changes in Mg and Ca concentrations mainly depend on the residence of water in carbonate systems and that is controlled by the distance from the recharge area and the dissolution and/or precipitation reaction of calcite and dolomite (Langmuir, 1971). The Mg/Ca ratio is commonly used in carbonate aquifers as an indicator of the residence time of circulating groundwater. The Mg/Ca ratios of

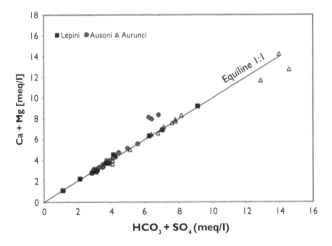

Figure 7.4 Relationship between Ca + Mg vs. HCO_3 + SO_4 in meq/l.

Table 7.2 Some ionic ratios. Total hardness and discharge elevations of springs (LP: Lepini. AS: Ausoni. AR: Aurunci).

Samples	Discharge elevations (m a.s.l.)	Mg/Ca	Na/Cl	Hardness (as CaCO₃) (mg/l)	Indication
LP01	5	0.68	1.00	318.58	Very hard
LP02	42	0.38	1.32	228.16	Hard
LP03	64	0.08	1.11	221.68	Hard
LP04	185	0.28	1.18	218.44	Hard
LP05	12	0.66	1.01	461.24	Very hard
LP06	840	0.21	1.34	157.76	Hard
LP07	5	0.60	1.15	348.35	Very hard
LP08	1110	0.14	1.08	140.98	Moderately hard
LP09	740	0.04	1.15	153.84	Hard
LP10	360	0.31	0.88	112.16	Moderately hard
LP11	360	0.46	0.82	56.16	Soft
LP12	1065	0.06	0.99	186.41	Hard
AS01	92	0.24	0.80	191.14	Hard
AS02	98	0.25	0.70	187.96	Hard
AS03	95	0.18	0.63	180.83	Hard
AS04	95	0.24	1.01	144.70	Moderately hard
AS05	6	0.55	0.90	238.41	Hard
AS06	4	0.65	0.90	259.42	Hard
AS07	2	0.88	0.88	419.73	Very hard
AS08	10	0.87	0.88	499.52	Very hard
AS09	6	0.85	0.86	407.91	Very hard
AS10	4	0.88	0.93	280.88	Very hard
AS11	18	0.20	0.73	157.70	Hard
AS12	20	0.23	0.75	167.12	Hard
AS13	10	0.62	1.57	167.45	Hard
AS14	475	0.08	1.04	199.30	Hard
AS15	100	0.14	0.94	189.55	Hard
AS16	500	0.11	0.93	211.65	Hard
AR01	18	1.51	1.64	639.02	Very hard
AR02	7	1.01	0.99	332.31	Very hard
AR03	13	1.12	1.10	347.49	Very hard
AR04	15	0.94	1.14	381.39	Very hard
AR05	17	0.37	2.85	413.41	Very hard
AR06	12	1.03	1.16	326.06	Very hard
AR07	14	1.34	1.06	361.56	Very hard
AR08	13	1.46	1.68	585.36	Very hard
AR09	8	0.75	1.15	396.60	Very hard
AR10	12	1.16	1.17	388.69	Very hard
AR11	39	0.29	0.79	175.10	Hard
AR12	31	0.46	1.74	183.18	Very hard
AR13	85	0.24	0.88	194.37	Very hard
AR14	77	0.24	0.93	188.56	Very hard

(Continued)

Table 7.2 (Continued).

Samples	Discharge elevations (m a.s.l.)	Mg/Ca	Na/Cl	Hardness (as CaCO₃) (mg/l)	Indication
AR15	83	0.20	0.84	252.16	Very hard
AR16	70	0.24	0.91	192.70	Very hard
AR17	74	0.20	0.90	202.48	Very hard
AR18	17	0.34	0.88	149.33	Moderately hard
AR19	17	0.29	0.86	151.95	Hard
AR20	11	0.45	1.50	712.71	Very hard
AR21	65	0.32	2.01	150.86	Hard
AR22	1	0.28	0.82	194.45	Very hard
AR23	1050	0.05	2.12	148.18	Hard
AR24	700	0.05	0.94	158.71	Hard
AR25	870	0.04	1.13	188.38	Very hard
AR26	700	0.03	1.33	199.19	Very hard

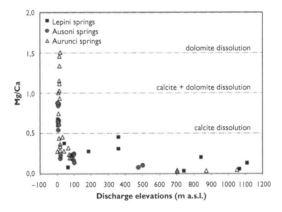

Figure 7.5 Mg/Ca ratio of spring samples vs. discharge elevations (m a.s.l.).

sampled waters are plotted against discharge elevations in Figure 7.5. The relationship between dolomite dissolution and calcite precipitation is thought to increase the Mg/Ca ratios along the flow paths. High elevation springs discharges near the recharge area and have the lowest Mg:/Ca ratios (<0.1), while low-elevation springs farther from the recharge area show higher Mg/Ca ratios (up to 1.5). In general, an increase water temperature can accelerate the kinetics of the dissolution of dolomite, and hence also the Mg/Ca ratio (Herman & White, 1985). The highest Mg/Ca ratios (~1.5) are found in the high temperature Aurunci springs, discharging at lower elevations, highlighting long residence time and enhanced weathering along the groundwater flow paths.

7.4.2   Effects of Ca and Mg on drinking water quality

The presence of low Ca and Mg concentrations in drinking water causes adverse health effects, however, high concentrations may cause domestic and industrial problems (Yang *et al.*, 2000; Reynolds & Richards, 1996). According to previous studies, the minimum required amounts of Ca and Mg in drinking water are 20 and 10 mg/l respectively, and the desired amounts of M and Ca in drinking water are 30–20 mg/l and 80–40 mg/l, respectively (Kozisek, 2003; Cotruvo, 2009). The maximum and minimum values of Ca and Mg concentrations are present in Table 7.1. The calcium concentrations in water samples range from 15.40 to 197.3 mg/l with minimum and maximum values, respectively. Almost 33.3% of the samples contain Ca concentrations higher than 75 mg/l, while about 66.3% of the springs show Ca concentrations below 75 mg/l. Mg concentrations range between 1.4 to 93.4 mg/l. Most of the samples (~66%) show Mg concentrations <30 mg/l. However, about 9.2% of the 54 samples have a Mg concentration >50 mg/l. The remaining water samples have magnesium concentrations within the range of 30–50 mg/l. The highest calcium (197.3 mg/l) and magnesium (93.4 mg/l) concentrations were observed in water samples from Aurunci springs.

The relationship between Electrical Conductivity (EC) and the Ca–Mg distributions is shown in Figure 7.6. Electrical conductivity values of spring water samples ranges from 138 to 2310 µS/cm indicating fresh (<500 µS/cm), marginal (500–1500 µS/cm) and brackish water types (>1500 µS/cm) in the area. Most of the samples have EC values less than the maximum permissible limit. The greater the conductivity, the greater is its salt content. Ca and Mg contents decreases gradually in groundwater samples, taken near coastal area, with increasing EC. The highest Ca and Mg concentrations were found for water samples with EC of 500–1000 µS/cm representing a typical water from the carbonate aquifer. The decrease in the Ca with increase in EC is possibly due to the proximity of sampling locations to the coast (i.e. samples discharged at lower elevations). The presence of higher concentrations is

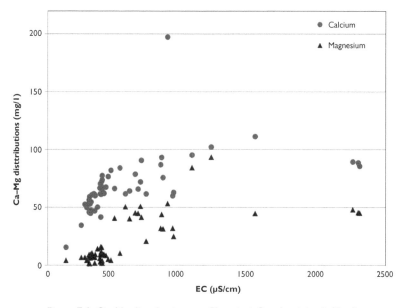

Figure 7.6 Ca–Mg distributions vs. Electrical Conductivity (µS/cm).

attributed to the influence of seawater in the coastal area and weathering of Mg-rich dolomite and calcareous-dolomitic lithologies.

A and Mg are the most important constituents of drinking water, moreover they are major contributors to water hardness. As contributors to hardness, calcium and magnesium ions can affect drinking water quality. High levels of total hardness do not cause a health risk, however, but both the extreme degrees, very soft (<75 mg/l as $CaCO_3$) and very hard (>300 mg/l as $CaCO_3$), are considered as undesirable features in water. Hardness levels between 80 and 100 mg/l (as $CaCO_3$) are generally acceptable in drinking water and are considered tolerable by consumers (Ternan, 1972; USGS, 2012; Bernardi et al., 1995; Memon, 2011). Besides, higher Ca and Mg concentrations affect aesthetic parameters such as taste, smell and colour of water (Lenntech, 1998). Determination of water hardness is a useful test to measure quality of water for drinking, agricultural and industrial uses. The total hardness of water is the sum of calcium and magnesium hardness expressed as mg/l $CaCO_3$. The total hardness (as $CaCO_3$) of water samples can be calculated by the following equation (WHO, 2008):

$$\left[CaCO_3\right] = 2.5\left[Ca^{2+}\right] + 4.1\left[Mg^{2+}\right]$$

The US-EPA classified water that contains 0 to 75 mg/l $CaCO_3$ as soft, 75 to 100 mg/l $CaCO_3$ as moderately hard, 150 to 300 mg/l $CaCO_3$ as hard and >300 mg/l $CaCO_3$ as very hard. The total hardness of Lepini spring samples range from 56 to 461 mg/l (Table 7.2) and fall between soft and very hard water category. For the spring water samples from the Ausoni Mountain total hardness range between 144 to 499 mg/l and is classified as moderately hard to very hard water. The highest total hardness values were observed in water samples from Aurunci Mountains ranging from 148 to 712 mg/l, respectively. Almost all the Aurunci spring samples are characterised as very hard water. The classification of water based on total hardness shows that most of the spring water samples fall between hard and very hard water type. Hardness levels between 80 and 100 mg/l (as $CaCO_3$) are recommended by World Health Organisation (WHO) as the minimum guideline limit for total hardness in drinking water. Waters with hardness levels in excess of 200 mg/l are considered poor but have been tolerated by consumers, however waters with hardness in excess of 500 mg/l are not suitable for most domestic purposes. Almost all sampled spring waters exceed the minimum allowable limit. The observed high total hardness values in spring water samples are related to the main rock types in the area investigated, where limestone, dolomitic limestones and dolomites are the most dominant formations.

7.5 CONCLUSIONS

Hydrochemical data and geochemical modelling techniques were employed to identify the main processes responsible for the evolution of spring waters from the carbonate aquifers of southern Latium region, Central Italy. The variations in major ion concentrations and hydrochemical facies suggest that these water types undergo further geochemical evolution through water-rock interaction along the flow paths with some seawater contamination near the coast. The classification of water based on total hardness (as $CaCO_3$) shows that most of the spring water samples fall between the hard (150–300 mg/l) and very hard (>300 mg/l) water types. The observed high total hardness in spring water samples are mainly related to the main rock types in the area, where limestone, dolomitic limestones and dolomites are common. The high dissolution rate for carbonate formations and the residence times of the circulating water in carbonate rocks increase the Ca and Mg concentrations and hence also the total hardness. Almost all sampled spring waters exceed the minimum guideline limit for drinking water. Hard water has no adverse effects on human health, however, it may create some problems for domestic and industrial users. The highest Ca and Mg concentrations were found in water samples having EC levels from 500 to 1000 µS/cm representing a typical water from the carbonate aquifer. The results of hydrochemical analyses and the calculated parameters (i.e. total hardness, TDS, pH, EC) suggest that most of the groundwater samples are suitable for drinking purposes. The observed parameters fall within the recommended limits of U.S. Environmental Protection Agency (US-EPA), World Health Organisation (WHO).

ACKNOWLEDGEMENTS

The authors would like to thank the Regional Basins Authority of Latium for the financial support of the project.

REFERENCES

Accordi B., Biasini A., Caputo C., Devoto G., Funiciello R., La Monica G.B., Palmieri E.L., Matteucci R. & Pieruccini U. (1976) Geologia e dissesti del territorio montano della Regione Lazio, In: *Carta della Montagna 2, Monografia Regionali No 12 Lazio*, Ministero di Agricoltura, Roma, 55–101.

Bernardi D., Dini F.L., Azzarelli A., Giaconi A., Volterrani C. & Lunardi M. (1995) Sudden cardiac death rate in an area characterised by high incidence of coronary artery disease and low hardness of drinking water. *Angiology* 46, 145–149.

Boni C. (1975) The relationship between the geology and hydrology of the Latium-Abruzzi Apennines. In: M Parotto, A Praturlon, Geological summary of the central Apennines, *Quaderni de La ricerca scientifica* 90, 301–311.

Briz Kishore B.H. & Murali G. (1992) Factor analysis for revealing hydrochemical characteristics of a watershed. *Environmental Geology* 19, 3–9.

Celico P. (1978) Schema idrogeologico dell'Appennino carbonatico centro-meridionale. *Memorie e Note dell'Istituto di Geologia Applicata* 14, 1–97.

Cerling T.E., Pederson B.L. & Damm K.L.V. (1989) Sodium-calcium ion exchange in the weathering of shales: Implications for global weathering budgets. *Geology* 7, 552–554.

Chenini I. & Khmiri S. (2009) Evaluation of ground water quality using multiple linear regression and structural equation modeling. *International Journal of Environmental Science and Technology* 6(3), 509–519.

Cotruvo J. & Bartram J. (2009) *Calcium and magnesium in drinking-water: Public health significance.* Geneva, World Health Organisation, WHO.

Ettazarini S. (2005) Processes of water-rock interaction in the Turonian aquifer of Oum Er-Rabia Basin, Morocco. *Environmental Geology* 49, 293–299.

Ferrandiz J., Abellan J.J., Gomez-Rubio V., López-Quílez A., Sanmartín P., Abellán C., Martínez-Beneito M.A., Inmaculada M., Vanaclocha H., Zurriaga O., Ballester F., Gil J.M., Pérez-Hoyos S. & Ocaña R. (2004) Spatial analysis of the relationship between mortality from cardiovascular and cerebrovascular disease and drinking water hardness. *Environmental Health Perspectives* 112(9), 1037–44.

Fisher R.S. & Mullican W.F. (1997) Hydrochemical evolution of sodium-sulfate and sodium-chloride groundwater beneath the Northern Chihuahua Desert, Trans-Pecos, Texas, USA. *Hydrogeology Journal* 52, 4–16.

Herman J.S. & White W.B. (1985) Dissolution kinetics of dolomite: effects of lithology and fluid flow velocity, *Geochimica et Cosmochimica Acta* 49, 2017–2026.

Jawad S.B. & Hussien K.A. (1986) Contribution to the study of temporal variations in the chemistry of spring water in karstified carbonate rocks. *Hydrological Sciences – Journal – des Sciences Hydrologiques* 31(4), 529–541.

Johnson R. & Scherer T. (2012) Drinking water Quality: Testing and Interpreting Your Results North Dakota State University, WQ-1341.

Kozisek F. (2003) Health significance of drinking water calcium and magnesium National Institute of Public Health Prague; Czech Republic. www.midasspringwater.com/typed%20 documents/HealthSignificance.pdf.

Langmuir D. (1971) Geochemistry of some carbonate ground waters in Central Pennsylvania. *Geochimica et Cosmochimica Acta* 35(10), 1023–1045.

Lenntech (1998) www.lenntech.com/ro/water-hardness.

McLean W., Jankowski J. & Lavitt N. (2000) Groundwater quality and sustainability in an alluvial aquifer, Australia. In: Sililo O *et al.*, eds. *Groundwater past achievements and future challenges*, A Balkema, Rotterdam, 567–573.

Memon M., Soomro M.S., Akhtar M.S. & Memon K.S. (2011) Drinking water quality assessment in Southern Sindh, Pakistan. *Environmental Monitoring and Assessment* 177, 39–50.

Parkhurst D.L. & Appello C.A.J. (1999) *User's guide to PHREEQC., version 2. – A computer program for speciation, batch-reaction, one-dimensional transport, and inverse geochemical calculations.* US Geological Survey Water-Resources Investigations Report 99-4259, 312 p.

Plummer L.N. (1977) Defining reactions and mass transfer in part of the Floridan aquifer. *Water Resources Research* 13(5), 801–812.

Reynolds, T.D. & Richards, P.A. (1996) *Unit operations and processes in environmental engineering.* PWS Publishing Company, Boston, MA.

Rubenowitz E., Axelsson G. & Rylander R. (1996) Magnesium in drinking water and death from acute myocardial infarction. *American Journal of Epidemiology* 143(5), 456–462.

Sappa G., Barbieri M., Ergul S. & Ferranti F. (2012) Hydrogeological conceptual model of groundwater from carbonate aquifers using environmental isotopes (18O, 2H) and chemical tracers: A case study in southern Latium region, Central Italy. *Journal of Water Resource and Protection (JWARP)* 4, 695–716.

Shuster E.T. & White W.B. (1971) Seasonal fluctuations in the chemistry of limestone spring A possible means for characterizing carbonate aquifers. *Journal of Hydrology* 14, 93–128.

Stumm W. & Morgan J.J. (1996) *Chemical equilibria and rates in natural waters, Aquatic Chemistry.* John Wiley and Sons, New York, 1022 p.

Ternan J.L. (1972) Comments on the use of a calcium hardness variability index in the study of carbonate aquifers: with reference to the central Pennines, England. *Journal of Hydrology* 16, 317–321.

US Environmental Protection Agency (1986) *Quality criteria for water 1986.* EPA 440/5-86-001. Washington, DC 20460.

USGS (2012) http://water.usgs.gov/owq/hardness-alkalinity.html#hardness.

Wen X.H., Wu Y.Q. & Wu J. (2008) Hydrochemical characteristics of groundwater in the Zhangye Basin, Northwestern China. *Environmental Geology* 55(8), 1713–1724.

White W.B. (1988) *Geomorphology and Hydrology of Karst Terrains*. Oxford University Press, New York, 464 p.

WHO (2004) *Guidelines for Drinking water Quality*, 3rd edn. *Vol 1 Recommendations*, World Health Organization, Geneva.

WHO (2008) *Guidelines for Drinking-water Quality*, 3rd edn. *Incorporating The First And Second Addenda Volume 1 Recommendations*. World Health Organization, Geneva.

Yang C.H., Chiu H.F., Cheng M.F., Hsu T.Y., Cheng M.F. & Wu T.N. (2000) Calcium and magnesium in drinking water and the risk of death from breast cancer. *Journal of Toxicology and Environmental Health* 60, 231–241.

Yang C.Y. & Hung C.F. (1998) Colon cancer mortality and total hardness levels in Taiwan's drinking water. *Archives of Environmental Contamination and Toxicology* 35(1), 148–51.

Chapter 8

The major litho-structural units in selected areas of Northern Nigeria: Some statistics on the distribution of Ca and Mg are appended

Hyeladi Usman Dibal & Uriah Alexander Lar
Department of Geology and Mining, University of Jos, Jos, Nigeria

ABSTRACT

In Northern Nigeria, four major litho-structural units constitute the main aquifers from which water is abstracted for domestic, industrial and agricultural purposes; the Basement Complex, Sedimentary formations, Younger Granites, and the Volcanics. In general, the Basement Complex units are calc-alkaline displaying a composition that varies from Fe/Mg rich to Na/K rich. The basic and intermediate rock components of these units are Ca/Fe rich with low MgO content. In general, the yields in the Basement Complex terrain are good depending on the degree of weathering of the rock, whereas in the Younger Granites province, the fractured crystalline aquifer has higher yields than the soft overburden and volcanic aquifer. The sedimentary formations are predominately clastic materials intercalated with marine sediments and carbonates. The (Bima) Sandstones are good water reservoirs whose yields depend on their degree of fracturing. Some statistical data on the distribution of Ca and Mg ions is appended.

8.1 GEOLOGY

Northern Nigeria has an aerial extent of 692 768 km² defined by latitudes 5° and 15° N and longitudes 3°15′ and 15′ E. Northern Nigeria is bounded to the north by Chad and Niger Republic, to the east by northern Cameroon, to the west by northern Benin Republic and to the south by the southern states of Oyo, Ekiti, Edo, Delta, Enugu and Cross River states. Northern Nigeria contains nineteen of the 36 states in Nigeria (Figure 8.1).

In Northern Nigeria four major litho-structural units constitute the main aquifers from which water is abstracted for domestic, industrial and agricultural purposes. These are basement, sedimentary, younger granites, and the volcanic aquifers. The basement and sedimentary units constitute over 95% of the rock units and the remaining 5% comprises the younger granites and the volcanic units. The basement units are Precambrian in age, and consist of three different lithologies: migmatite-gneiss, meta-sedimentary and meta-volcanic rocks (Schist Belts) and Pan-African Granitoids (Older Granites). The sedimentary lithologic units are Cretaceous to Quaternary; the younger granites are Jurassic, while the volcanic lithologies are Tertiary in age.

The Migmatite-Gneiss Complex consists of grey foliated biotite and/or biotite hornblende quartzo-feldspathic gneiss of tonalitic to granodioritic composition (Rahaman, 1981). The presence of mafic to ultramafic outcrops have often been reported as discontinuous, boudinage lenses or concordant sheets of amphibolites

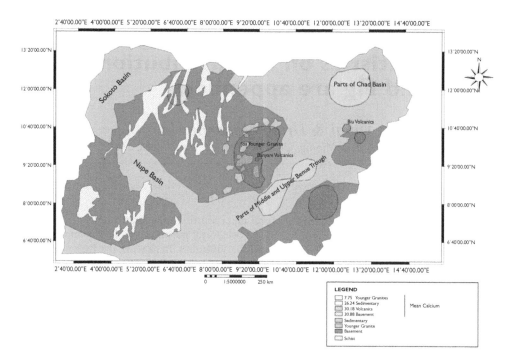

Figure 8.1 A simplified geological map of Nigeria.

with minor amounts of biotite-rich or biotite-hornblende-rich ultramafic. There exist also felsic components of varied group of rocks consisting essentially of pegmatite, aplitites, quartz-oligoclase veins, fine-grained granites, granite gneiss and porphyritic granites. The schist belt is generally referred to as metasedimentary and meta-volcanic rocks which are largely sediment-dominated and composed essentially of pelites, semi-pelites and quartzites. Amphibolites and ultramafic rocks occur in some places. They also host important marbles and Banded Iron Formation mineralisation (Russ, 1957; Truswell & Cope, 1963). The older granites (Pan African granites) refer to several important petrologic groups of the same age and include biotite and biotite-muscovite granites, syenites, charnockites, diorites. Monzonites, serpentinites and anorthosites. In general, the basement rocks contain variable proportions of plagioclase (>60%), quartz, and biotite. The older granites are calc-alkaline displaying composition that varies from Fe + Mg-rich to Na + K-rich. The basic and intermediate rock components such as diorite consist of 40 to 60% of andesine and are Fe-rich with low MgO content. Orthoclase, hypersthenes, biotite, ilmenite and magnetite are present in varying proportions. In the crystalline basement terrains, the availability of water depends on the degree of fractures or weathering of the rock. In general, the water yields in the basement complex terrain are favourable to good where the rock is severely fractured.

The Younger Granites display a variety of rock types, with similar petrographic characteristics which include; biotite granite, hornblende biotite granites, rhyolite, riebeckite–biotite granite, syenite, gabbro and pyroxene-fayalite granite (Macleod *et al.*, 1971). Volcanic rocks form minor components in the host granitic bodies. Most of the volcanic rocks have been altered into lateritic clays, gibbsitic clays and calcite in places (Lar *et al.*, 2000). The basaltic volcanics are tholeiitic (Fe + Mg-rich). The albite riebeckite granites exhibit some degree of alteration and are enriched in Cl, F, Li, Zr and Zn. The pyroxenes and amphiboles in these granites are Ca + Na-rich with minor $Fe^{3+} > Al^{3+}$ and Fe^{2+} substitutes for Mg^{2+}. In the younger granites three aquifers have been identified; the fractured crystalline aquifer, soft overburden aquifer and volcanic aquifer (Schoeneich & Mbonu, 1991). In most cases, the fractured crystalline aquifer has good yields.

The sedimentary basins in northern Nigeria are principally that of the Benue Trough (Lower Benue, Middle Benue, and Upper Benue) and the Chad Basin. Other basins include the Sokoto and Bida basins. The Benue trough contains a 6 km thick Cretaceous infill sediment deposited over a shallow basement structure. The sediments were folded during the Campano-Santonian tectonic episode and subsequently overlain by the Kerri-Kerri Formation. The sediments are dominated by clastic materials of the deltaic Bima Formation interrupted locally by marginal marine sediments with some carbonates. Volcanic rocks and hyperbyssal mafic intrusions occur along the trough inter-bedded with sediments of different ages (Benkhelil, 1989). In the Chad basin, there are three layers of aquifers, the top is water table aquifer, the middle and the lower artesian aquifers. Recharge into the middle and lower aquifers is sustained by rainwater infiltrating from the more humid parts of the basin (Offodile, 1976). In the Benue trough, the Bima Sandstones are well cemented and a good water reservoir. Water yield in these sandstones depends on the degree of fracturing and weathering (Offodile, 1976; Ishaku & Ezeigbo, 2000). The Yolde Formation has proven to be a more reliable water reservoir than the Bima Sandstones.

8.2 GEOLOGY OF THE MAJOR LITHO-STRUCTURAL UNITS IN NIGERIA

The surface area of Nigeria is underlain in nearly equal proportions by crystalline and sedimentary rocks (Figure 8.2).

The crystalline rocks are further divided into three main groups:

1 The Basement Complex
2 The Younger Granites
3 The Tertiary to Recent Volcanic.

8.2.1 The Basement Complex

The Basement Complex outcrops are distributed in two main areas: a roughly circular area in north central Nigeria, and a rectangular area broken up into three zones by sedimentary rocks on the eastern border with Cameroon Republic and Benin Republic.

The geology of the Nigerian Basement Complex has been described by Russ (1957); Truswel & Cope (1963) in the north west; and by Jones & Hockey (1964) Oyawoye (1964) in the south west. The works of McCurry (1971; 1973; 1976); Ogezi (1977); Grant (1978); Ajibade *et al.* (1979); Turner (1983) Fitches *et al.* (1985) in the

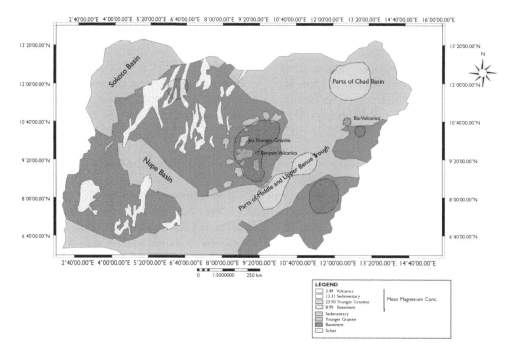

Figure 8.2 Mean Ca concentration (mg/l) in some parts of northern Nigeria.

north west allow some correlations to be made with work by Burke & Dewey (1972); Burke *et al.* (1976); Annor & Freeth (1985); Annor (1998) and Caby (1989) in the south west and confirm four distinct basement lithologies:

1   The Migmatite-Gneiss Complex
2   The Metasedimentary and Metavoleanic rocks (The Schist Belts)
3   The Pan-African Granitoids (The Older Granites) and
4   The Undeformed acid and basic dykes.

8.2.1.1 The Migmatite-Gneiss Complex

The Migmatite-Gneiss Complex is regarded as the Basement and it is the most widespread of the main units in the Basement Complex of Nigeria. It is a heterogeneous assemblage including migmatites, orthogenesis and a series of basic and ultra basic metamorphosed rocks (Rahaman, 1988). Petrographic evidence indicates that the Pan African reworking has led to re-crystallisation of many of the constituent minerals of the Migmatite Gneiss Complex by partial melting, with most displaying medium to upper amphibolite metamorphism. Three petrological units characterise the Migmatite-Gneiss Complex.

1   Grey foliated biotite and/or biotite hornblende quartzo-feldspathic gneiss of tonalitic to granodioritic composition, which is now known as the grey gneiss or early gneiss (Rahaman, 1981).

2   Mafic to ultramafic component, which were present, often outcrops as discontinuous, boudinage lenses or concordant sheets of amphibolites with minor amounts
of biotite-rich or biotite-hornblende-rich ultramafic. Except where it constitutes
the paleosome to the migmatite, it is present in subordinate amounts to the grey
gneiss.

3   Felsic component is a varied group of rocks consisting essentially of pegmatite,
aplitites, quartz-oligoclase veins, fine-grained granites, granite gneiss, porphyritic
granites. Other works done on the migmatite gneiss complex are those of Dada
et al. (1993; 1995) in the Kaduna area.

8.2.1.2   The schist belts (meta-sedimentary and metavolcanics rocks)

The metasedimentary and metavolcanic rocks generally called the schist belt, occupy
north-south trending synformal troughs infolded into the Older Migmatite-Gneiss
Complex and are better developed in the western half of the country. They are
largely sediment-dominated and the important lithologies are pelites, semi-pelites and
quartzites. Some belts host metamorphosed chemical sediments (marbles and Banded
Iron Formation while mafic to ultramafic rocks now present as amphibolites and
ultramafic rocks occur in others. Felsics to intermediate meta-volcanics and grey-
wackes have also been described. The belts are among the best-studied groups of
rocks in Nigeria (Russ, 1957; Truswell & Cope, 1963) because of the associated
mineralisation of gold, the Banded Iron Formation, and the occurrence of marble,
manganese, talc and anthophyllitic asbestos. The more recent studies, especially that
of Garba (2002) have emphasised two prominent regional north-north-east to south-
south-west Pan African wrench faults related to gold and associated mineralisation.

8.2.1.3   The Pan African granitoids (older granites)

The term older granites was introduced by Falconer (1911) on the basis of morphol-
ogy and texture that distinguished the Pan-African granitoids from the Carboniferous-
Cretaceous high level anorogenic volcanic, hyperbyssal, peralkaline, younger granites
of the Jos Plateau Region. The term Pan-African granitoids is to be preferred not only
because of its merit on age, which was not available at the time, but because it covers
several important petrologic groups formed at the same time. The granitoids include
biotite and biotite-muscovite granites, syenites, charnockites, diorites, monzonites,
serpentinites anorthosites. In many places, the coarse-grained biotite-hornblende granites have concordant foliation with the host Migmatite Gneiss Complex or schists.

8.2.2   Younger granites

The Younger Granites are characterised by circular intrusions and represent one of the
classical areas of occurrences of ring-complexes in the world. They are non-orogenic
rocks, which have intruded the Late Precambrian to Lower Paleozioc Basement
Complex of Northern Nigeria in an north south trend (Turner, 1983). Ring faulting
and cauldron subsidence are the major tectonic controls governing the emplacement
of the younger granite (Macleod *et al.*, 1971). These controls have operated during

the volcanic and the stages of the emplacement cycles. The pattern of initial volcanism has been determined largely by ring fracturing. These fractures controlled the alignment and distribution of the vents and the major cauldron of the same mechanism at greater depth, beneath the lava accumulation, which localised the peripheral ring dykes and the granitic plutons. The younger granites exhibit a variety of rock types, which present similar characteristics throughout the whole province. These rock types include; biotite granite, hornblende biotite granites, rhyolite, riebeckite–biotite granite, syenite, gabbro, pyroxene–fayalite granite (Macleod *et al.*, 1971).

8.2.3 Sedimentary basins

The sedimentary basin in northern Nigeria are referred to as the intra-continental basins the Benue Trough (Lower Benue, Middle Benue, Upper Benue) the Bida, the Sokoto and the Chad Basins.

8.2.3.1 Benue Basin

The Benue Trough is a linear north east–south west trending structure, about 800 km in length and has a maximum width of about 150 km. It opens southwards into the Gulf of Guinea with the Cenozioc Niger Delta. At its north-eastern end, it bifurcates, the north-south trending Gongola Basin and the east west trending Yola arm. The Benue trough is thus a complex of several sub-basins. Sedimentary infill in the areas flanking the axial zone of shallow basement rocks contain basic to intermediate intrusives, believed to reach a thickness of up to 6 km. Benkhelil (1989) has suggested that the features of the Benue trough complex can be explained by a pull-apart Basin model in which existing basement fractures were rejuvenated and wrenching along these fault systems resulted in block faulting and formation of the sub-basin. The sediments deposited in this basin were folded during the Campano-Santonian tectonic episode and subsequent sedimentation (Kerri-Kerri Formation) was restricted to the north western flank of the deformed sediments.

To the north and north east, the sediments are dominated by clastic material of the deltaic Bima Formation interrupted locally by marginal marine sediments with some carbonates. Volcanic rocks and pyroclastic and hyperbyssal mafic intrusions occur along the trough interbedded with sediments of different ages.

8.2.3.2 Chad Basin

The Chad Basin is the largest inland basin in Africa, occupying an area of approximately 2 500 000 km², and extending over parts of the Republic of Niger, Chad, Sudan, and the northern part of Cameroon and Nigeria. Like the Lullemmeden Basin, the Chad Basin has a history that goes back to Paleozoic age. In the Nigerian sector of the Chad Basin, no rocks of Paleozoic age occur, but the basin is covered by a blanket sequence of Quaternary sediments, the Chad Formation, comprising lacustrine clays with occasional sand horizons, dipping at low angles towards the north-eastern corner of the basin.

8.3 HYDROGEOLOGY OF THE MAJOR LITHO-STRUCTURAL UNITS IN NIGERIA

The hydrogeology of Nigeria reflects its geology and is categorised into the crystalline Basement Complex area and the sedimentary basins (Offodile, 1976).

8.3.1 Hydrogeology of the Basement Complex areas

Over 50% of northern Nigeria is covered by the rocks of the Basement complex. This extensive mass of almost impermeable rocks underlies many parts of the arid and semi-arid areas of the north, in which potable water is scarce.

8.3.1.1 Hydrogeology of the Older Granites and Migmatite-Gneiss Complex areas

The rocks include different textures of granites; coarse to fine consisting essentially of biotite, feldspar quartz, which are indications of the hydrogeological characteristics (Offodile, 1976). While the coarse granites weather into water bearing sandy residue, the syenitic rock types, with the predominance of the unstable minerals such as feldspars, decompose into plastic or soft clays and other argillites which behave only as aquitards or aquicludes.

Generally only small amount of water can be obtained in the freshly unweathered bedrock below the weathered layers. Even when fractured, the clayey materials tend to seal the openings of the fractures and prevent water from being transmitted to boreholes. Experience on the Nigerian Basement Complex rocks shows that yields of between 1 to 2 l/s are good. In highly fractured zones yields up to 4 l/s have been obtained. Yields less than 1 l/s are useful for hand pumps. However boreholes producing less than 0.5 l/s, are regarded as dry, even for hand pumps.

8.3.1.2 Hydrogeology of the Meta-sediments Quartzite and Schist Complex areas

The hydrogeological properties depend on the texture of the meta-sediments hence the schists and the phyllites are poor aquifers, while the quartzites and pegmatites are good aquifers where fractured.

8.3.1.3 Hydrogeology of the Younger Granites/Volcanic areas

Three aquifers exist in the Younger Granites/Fluvio Volcanic Province; the fractured crystalline aquifer, soft overburden and volcanic aquifer (Schoeneich & Mbonu, 1991). The fractured crystalline aquifer contains water in amounts in which open well and sometimes boreholes can be sited in tectonically fractured zones. The major fractured zones generally have high yielding boreholes usually from 3 l/s and above. Transmisivity value of this aquifer is between 0.58 to 1.08 m²/hr and hydraulic conductivity values range from 0.13 to 0.23 m/hr.

The soft overburden aquifer consists predominantly of clay and insitu chemically weathered rocks. Lithologically, it is highly variable. The most common constituents are sands and gritty clays formed as a result of the burrowing activities of termites. This makes the soft overburden aquifer lack natural filtration (Schoeneich & Mbonu, 1991). In the Biu volcanic area, three sources with high groundwater potential exist; the eluvium, the weathered and fractured zones of the basalt and the contact between the basalt and the basement. The eluvium covers almost the entire Biu Plateau. This layer has an average thickness of 10 m, although it shows spatial variability. It is the source of a water table aquifer tapped by hand dug wells in the area. The basalt stores water either within the weathered zone and or within the fractures. The weathered zone is between 15 and 20 m thick and can provide yields up to 1 l/s. At depths of 50–70 m heavily fissured and jointed basalt occurs. Boreholes tapping this zones yield between 1–1.5 l/s (Conred, 1978).

8.3.2   Hydrogeology of the sedimentary basin

8.3.2.1   Chad Basin

The Chad Basin includes the Hadejia-Yobe Basin, or the south west Chad Basin and the east Chad basin. The basin area spreads out from the north-eastern corner of the country into the southern fringes of the Sahara Desert in Niger, Cameroun and Sudan. The Nigerian part of the basin consists of about one third of the whole basin. There are three aquifer layers, the top water table, the middle and the lower artesian aquifers. Recharge into the basin is mainly from the southern outcrop areas of the aquifers with higher rainfall rather than the northern fringes. Recharge into the middle and lower aquifer is sustained by direct rainfall recharge from the more humid parts of the basin. The Upper zone aquifer has a depth of 105 m (Offodile, 1976). The aquifer may either be water table or semi-confined. Yields are in the range 2.5 to 30 l/s. The Middle zone aquifer is separated from the Upper zone aquifer by about 150 m of clays. These clays confine the underlying Middle zone aquifer which in turn is separated from the Lower by another 120 m thickness of clay and shale.

8.3.2.2   Benue Basin

The Benue Basin is a continuous elongated geological structure, conveniently sub-divided into three basins described as Upper, Middle and Lower Benue sub-basins. Hydrogeologically, separation into these basins may not be possible since the basin is seen as one extended complex hydrogeological structure comprising a number of isolated aquifer units within a predominantly semi-permeable to impermeable geological formations (Offodile, 1976). The Upper Benue Basin has an area of about 203 000 km². It is underlain by patches of basement rocks, a number of volcanic plugs, basaltic flows and sedimentary rocks of Cretaceous age. The Upper Benue Basin is separated from the Chad Basin by the Zambuk Ridge, which stretches in a north east to south west, running from Zambuk to the Biu Plateau. Two potential groundwater basin have been demarcated in the Upper Benue Basin (Offodile, 1976) the Gombe to the north and the Lau to the south separated by the Lamurde Anticline.

The sub-basins contain permeable formations, which include the Bima, Yolde, Pindiga and the Gulani sandstones. Extensive deposits of the Bima Sandstone occur north and south of the River Benue (Offodile, 1976). The Bima Sandstones are well cemented and present hydrogeological characteristics of a basement rock.

Secondary permeability is developed by fracturing, weathering and solution processes. It is not generally, a good reservoir due to its poor permeability. Water occurs mostly under water table conditions. Sometimes, some clay beds confine it. Yields in the Bima Sandstones are unpredictable and range from 2 l/s to 8 l/s. An average of 1 to 5 l/s has been recorded depending on the hydrogeological environment (Offodile, 1976; Ishaku & Ezeigbo, 2000). The aquifers of the Yolde Formation have given a better and more reliable result than those of the Bima Sandstones. Yields of up to 4 l/s have been obtained in Gombe and Numan areas. The water normally occurs under water table conditions. In some places, however, the clay beds provide confined artesian or sub-artesian conditions.

In the Lau Basin, the Dukul, Sekule, Jcssu, and Numanha Shales consists mainly of clays, shales and other argillaceous materials. The Dukul Formation contains thin limestone beds which alternate with the shales. Groundwater occurrence in these formations is limited and is, therefore, described as an aquiclude. The Pindiga, which overlies the Yolde Formation within the basin, consists of black shales, limestones and a number of interbedded sands. The sands form a confined aquifer within the formation. Thompson (1965) reported that the aquifers are limited in extent and yields are small. Yields of about 1.2 to 5.5 l/s have been reported in Gombe. The Gombe Sandstones are generally a poor aquifer. The dominant argillaceous materials reduce the permeability considerably and render the formation useless as an aquifer. There is no hydrogeological data on the Lamja Sandstones available, which is a lateral equivalent of the Gombe Sandstone.

The Keri Keri Formation outcrops both in the Upper Benue Basin and the Chad Basin. This formation is a sequence of fine-grained sandstones, clays, and silts with some thin coal bands (Offodile, 1976). The lithology changes both vertically and laterally, du Preeze and Barber (1965) reported that, the maximum proven thickness of the formation is about 200 m. However, Okafor (1982) estimated by gravity survey, a total thickness of nearly 200 m in the Gombe area. Despite the looseness and coarseness of the formation, this apparently highly permeable sandstone formation is known to be unpredictable, hydrogeologically. Much of the arenaceous beds are dry with little or no water in the top beds. Groundwater tends to be perhed by interbedded clays, shales and siltstones, hence the large variation in depth of the water tables ranging from 0 to 200 m. In Ako near Bauchi the water table is about 60 m. South east of Bauchi (Bara and Yuli), a water level of 165 rn was recorded (Offodile, 1976).

The Asu River group underlies the Awe Formation and is essentially an aquitard or aquiclude. The overlying Awe Formation comprises flaggy, whitish, medium-coarse, sometimes calcareous sandstones, with interbedded shales, thin limestone and clays from which issue brines. Towards the base the sandstones become finer grained and more micaceous. The outcrops cover most areas of Awe, Zurak. Ribi and pats of Markudi. The sandstones beds are usually multi-layered and are highly porous and water yielding. However, the water is contaminated by the brines from the interbedded shales (Offodile, 1976). The shales act as confining beds. The Markudi Formation, like the Bima comprises highly indurated sandstones which are almost impermeable in some places. Where it is fractured or where less indurated it is more permeable and is a good aquifer. Yields up to 8.2 l/s have been obtained in the Markudi Formation. The Keana and the Ezeaku Formations in the Middle Benue Basin appear to be the most important aquifers. The Keana Sandstones usefulness as a potential groundwater reservoir depends on its secondary permeability derived also from weathering and fracturing.

The Agwu Formation in the Middle Benue consists of grey bedded shales with occasional sandstones and limestone beds. The sandstone beds are usually coarse-grained. Where coarse, they are very permeable and water bearing but often limited in thickness and lateral extent, hence reducing the groundwater potential. The overlying clay beds confine the lenticular sandy aquifers. Recharge is, therefore, restricted by the clay shale aquicludes. When the rocks are fractured or faulted the aquifers are interconnected and recharge into them increases considerably.

The basal sandstones of the Lafia Formation comprises essentially sandstones. This formation overlies the Agwu Formation. The sandstones are generally brownish at the top and whitish at the bottom. It is fine- to coarse-grained, friable and feldspathic. It ranges in thickness from about 10 to 150 m in Lafia area and thickens towards the south west. The formation thins out from the south west to the north east, forming a wedge against the older formation. The formation is highly permeable and gives rise to several springs at its contact with the less permeable Agwu Formation.

8.4   STATISTICAL DISTRIBUTION OF Ca AND Mg

Data from field work as well as from the literature were collated. Some of the data, especially those of Barber (1965) are over fifty years old. All other data are less than 15 years old. The cation and anion balance was calculated with AqQa software for some of the data and a large proportion of the data fall within the ±5% acceptable limit, suggesting that most of the data are credible. Most of the water samples were analyzed using the combined ICP AES/ICP MS techniques at both the Activation and ACME Laboratories in Canada. The analysis of other samples was carried out using the ICP OES at the Department of Geology and Mining, University of Jos, Nigeria.

Concentrations of Ca and Mg in some of the aquifer units in northern Nigeria as well as their minimum, maximum and mean values are presented in Tables 8.1–8.11. The mean values of these ions are also presented graphically in Figures 8.2 and 8.3. The Basement Aquifers have Ca concentration ranging from 1 to 119 mg/l and Mg from 3 to 154 mg/l. The Sedimentary Aquifers have Ca in the range of 3 to 154.7 mg/l and Mg between 0.60 to 55.90 mg/l. In the north-east sedimentary areas, where groundwater occurs in the Upper, Middle and Lower Aquifers, the Ca concentration in the Upper Aquifer ranges from 18.4 to 30.4 mg/l and Mg from 7.1 to 8.2 mg/l. The Middle Aquifer has Ca ranging from 3.40 to 63.6 mg/l and Mg from 0.1 to 27.70 mg/l. The Younger Granites Aquifers have Ca varying from 0.29 to 58.7 mg/l and Mg from 0.1 to 18.0 mg/l. The Volcanic Aquifers have Ca in the range

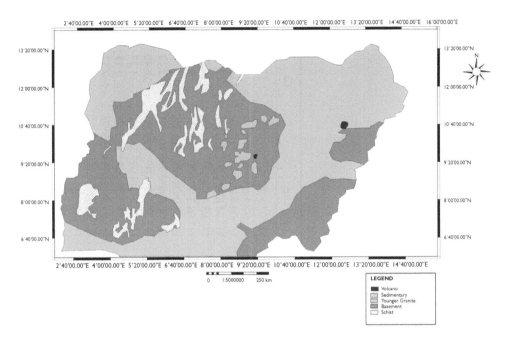

Figure 8.3 Mean Mg concentration (mg/l) in some parts of northern Nigeria.

of 1.0 to 31.5 mg/l and Mg 2.2 to 27.6 mg/l. 60% of the values of Ca in the basement aquifers are within the range of 0.11 to 29 mg/l and 42% are within the range 10 to 19 mg/l. 61% of Mg values are within the range of 0.1 to 9 mg/l. The sedimentary aquifers have 51% Ca concentrations within the range 0.5 to 30 mg/l and 64% of the Mg concentrations are within the range 0.1–15 mg/l. However, the majority of these values fall within the range 0.1 to 8.1 mg/l. These values are recorded in the Chad Basin Sedimentary Aquifers. The younger granite aquifers generally have lower Ca and Mg concentrations with over 80% of both ions below 10 mg/l. In the Volcanic Aquifer, very few water points show complete absence of both Ca and Mg, but higher values are obtained in the Biu Volcanic Plateau.

Over 90% of the samples in all the aquifers have both Ca and Mg below both the minimum and permissible limits of the standards of purity for drinking water given by Northern Nigeria Ministry of Health and the World Health Organisation.

The distribution of calcium and magnesium in the groundwaters of in some parts of Northern Nigeria indicate that the volcanic rock aquifer generally has a higher concentration of Ca and Mg than the other aquifers. The younger granite rock aquifer has a lower concentration of Ca and Mg compared to the other aquifers. The main source of Ca and Mg in the crystalline aquifers is the dissolution of ferromagnesian silicates and feldspars. In the sedimentary basins, the source of Ca is from the dissolution of calcite and feldspars while Mg derives from the dissolution of dolomite, gypsum and magnesite.

Table 8.1 Contents of Ca and Mg (mg/l) in groundwater from the Basement Rock Aquifers.

Coordinates	Ca	Mg	Coordinates	Ca	Mg	Coordinates	Ca	Mg
9° 11.309N 9° 46.724E	27.16	5.2	9° 06.506N 9° 47.410E	57.42	14.98	11° 07.518N 09° 31.242	22.70	7.03
9° 10.664N 9° 45.276E	16.67	3.5	9° 07.380N 9° 47.515E	37.65	2.99	11° 05.453N 09° 34.473E	11.34	5.09
9° 10.625N 9° 45.275E	25.55	2.22	07° 26.49N 11° 21.41E	5.4	2.00	11° 04.528N 09° 33.725E	28.56	9.55
9° 08.377N 9° 46.887E		2.85	09° 05.063N 09° 54.238E	16.4	0.80	11° 03.828N 09° 33.321E	18.94	7.88
9° 09.394N 9° 47.185E	8.96	1.17	09° 04.320N 09° 53.410E	16.9	1.10	11° 08.400N 09° 34.173E	10.01	3.82
9° 09.159N 9° 47.080E	25.8	2.58	09° 42.59N 11° 43.59E	10.8	9.90	11° 08.377N 09° 32.460E	5.84	1.55
9° 10.060N 9° 46.800E	12.88	2.09		21.7	2.50	11° 07.093N 09° 32.556E	6.66	0.09
9° 09.216N 9° 50.030E	57.58	14.61	07° 48.06N 11° 13.50E	15.2	2.30	11° 06.856N 09° 31.537E	3.12	0.38
9° 10.416N 9° 50.687E	51.58	13.84	07° 48.06N 11° 13.50E	11.3	2.70	11° 07.347N 09° 30.802	26.40	9.06
9° 11.017N 9° 50.551E	37.19	11.67	08° 54.40N 11° 21.14E	29.1	27.50	11° 07.541N 09° 33.444E	11.39	4.62
9° 10.914N 9° 49.486E	15.09	12.34	09° 11.651N 09° 53.270E	22.9	12.70	11° 04.728N 09° 33.587	1.11	0.67
9° 09.540N 9° 48.027E	50.39	19.96	09° 07.917N 09° 46.816E	27	12.40	11° 04.162N 09° 33.597E	7.52	2.83
9° 10.234N 9° 47.880E	20.39	6.5	09° 13.448N 09° 49.489E	49.7	10.20	11° 04.097N 09° 33.311E	1.71	0.57
9° 10.907N 9° 48.088E	31.01	12.27	10° 09.593N 12° 55.056E	109	13.90	11° 05.929N 09° 33.257E	0.13	0.34
9° 09.431N 9° 47.205E	36.21	7.97	10° 09.240N 12° 54.463E	14.7	1.60	11° 07.471N 09° 31.571E	1.34	2.54
9° 07.832N 9° 48.429E	117.1	28.12	10° 09.077N 12° 54.320E	18	2.10	10° 01.37N 09° 20.31E	41.71	10.32
9° 06.508N 9° 48.401E	35.09	7.89	10° 09.616N 12° 56.440E	14.9	2.60	10° 01.57N 09° 30.31E	14.23	3.91
9° 06.526N 9° 47.895E	51.16	36.48	10° 09.477N 12° 56.463E	14.9	1.30	10° 01.34N 09° 30.45E	104.62	22.04

9° 07.348N 9° 49.973	6.86	2.48	10° 09.329N 12° 56.593E	14	1.10	10° 01.25N 09° 31.40E	32.34	16.63
9° 05.759N 9° 48.609E	8.47	2.93	10° 09.298N 12° 56.681E	26.7	2.80	10° 01.28N 09° 32.05E	47.63	13.73
9° 08.502N 9° 48.127E	33.01	35.25	10° 09.298N 12° 57.990E	27.2	3.80	10° 01.49N 09° 33.13E	15.19	1.54
9° 05.665N 9° 50.664E	34.61	6.55	10° 09.428N 12° 57.731E	15.2	1.60	10° 01.40N 09° 37.29E	5.62	1.87
9° 06.496N 9° 52.473E	22.83	5.55	10° 24.819N 12° 44.496 E	14.3	1.60	10° 01.56N 09° 33.08E	41.13	9.41
9° 07.000N 9° 50.617E	25.26	9.11	10° 24.768N 12° 44.594E	33.1	6.90	10° 01.46N 09° 33.17E	13.23	4.62
9° 07.352N 9° 50.564E	26.49	5.21	10° 23.786N 12° 44.749E	69	14.20	10° 01.45N 09° 33.19E	16.68	3.74
9° 08.469N 9° 52.226E	35.43	7.03	10° 23.196N 12° 53.972E	18.9	2.60	10° 01.08N 09° 34.08E	48.23	15.31
9° 09.674N 9° 48.905E	12.64	35.42	11° 46.205N 07° 12.108E	87.8	17.20	10° 00.45N 09° 33.30E	18.45	5.62
9° 05.108N 9° 48.811E	7.71	17.91	11° 47.147N 07° 11.269E	59.3	15.40	10° 00.43N 09° 31.59E	119.61	28.90
9° 04.716N 9° 48.694E	10.85	2.59	11° 46.480N 07° 09.553E	14.3	1.60	10° 00.13N 09° 31.00E	38.73	18.04
9° 03.021N 9° 49.314E	5.91	28.44	11° 47.448N 07° 11.059E	85.7	14.70	10° 00.32N 09° 31.42E	46.51	7.41
9° 03.325N 9° 49.078E	25.35	4.4	11° 48.128N 07° 11.277E	59.2	17.20	10° 03.41N 09° 37.16E	36.04	2.31
9° 06.025N 9° 45.502E	39	4.82	12° 14.582N 06° 02.456E	19.4	15.40	10° 03.37N 09° 37.57E	48.34	8.15
9° 06.265N 9° 46.478E	23.12	17.18	12° 30.330N 06° 24.432E	70.3	6.90	10° 03.43N 09° 37.54E	14.69	5.74
9° 07.278N 9° 47.189E	48.33	7.93	12° 22.281N 06° 28.718E	89.6	12.40	10° 03.59N 09° 37.40E	16.59	4.62
9° 07.273N 9° 47.186E	23.38	25.85	12° 10.577N 06° 03.205E	113	58.70	12° 27.244N 06°10.541E	21.4	6.00

Table 8.2 Maximum mean and standard deviation of Ca and Mg (mg/l) in waters of the Basement Aquifers.

	N	Minimum	Maximum	Mean	Std. Deviation
Ca	111	0.13	119.61	30.8844	26.85055
Mg	112	0.09	58.70	8.9974	9.53240
Valid N	111				

Table 8.3 Ca and Mg content (mg/l) in waters of some parts of the Upper and Middle Benue Trough.

Coordinates	Ca	Mg	Coordinates	Ca	Mg
08° 08.447N 08° 47.497E	101.00	16.90	08° 08.957N 09° 07.393E	50.86	11.98
08° 31.386N 08° 31.820E	30.00	0.60	08° 08.31N 08° 47.33E	65.15	51.32
08° 13.545N 08° 34.241E	31.70	12.00	08° 09.35N 08° 48.30E	116.01	21.18
08° 06.085N 09° 07.771E	30.90	13.00	08° 09.10N 08° 48.07E	27.38	20.80
08° 03.089N 09° 04.018E	35.10	8.50	08° 06.55N 08° 45.48E	13.65	5.24
08° 05.092N 09° 04.020E	38.79	16.45	08° 07.09N 08° 45.09E	12.79	4.94
08° 06.455N 09° 08.762E	10.31	4.48	08° 07.34N 08° 45.59E	9.28	3.50
08° 06.700N 09° 09.426E	115.00	35.93	08° 06.20N 08° 45.36E	2.10	0.51
08° 06.686N 09° 10.570E	104.90	26.36	08° 06.40N 08° 45.09E	14.00	4.87
08° 06.664N 09° 10.664E	12.24	1.46	08° 02.01N 08° 46.10E	46.70	55.90
08° 06.664N 09° 09.150E	92.74	19.40	08° 02.11N 08° 46.12E	41.00	43.60
08° 06.716N 09° 08.709E	140.30	50.61	08° 10.13N 08° 46.06E	52.40	38.30
08° 06.717N 09° 08.020	82.80	33.32	–	29.96	16.42
08° 04.130N 09° 09.147	12.53	1.55	–	42.97	20.39
08° 06.138N 09° 08.730	18.41	3.16	–	88.50	43.20
08° 08.487N 09° 08.560	6.67	1.24	–	46.00	18.32
08° 06.646N 09° 08.780	16.69	7.92	–	7.33	2.87

Sources of Data: Dibal (2012); Mamidu (2012); Cyril (2010); Kehinde (2010); Oche (2010); Sola (2010).

Table 8.4 Minimum, maximum, mean and standard deviation for Ca an Mg (m/l) in waters of some parts of the upper and middle Benue Trough.

	N	Minimum	Maximum	Mean	Std. Deviation
Ca	34	2.10	140.30	45.4753	37.72282
Mg	34	0.51	55.90	18.1241	16.55177
Valid N (listwise)	34	–	–	–	–

Table 8.5 Ca and Mg content (mg/l) in waters in the Middle Zone Aquifer of the Chad Basin.

Dikwa 1	20.8	9.8	Wulo	9.6	4	Ngaratuwa	47.8	22.2
Shuari	24.8	18.9	Bulongowa	24	15.6	Meleram	51.2	16.3
Kesa Ngala	20.8	7.6	Dule Gana	8	8.2	Lerwi	56.8	23.8
Mafa 1	56.8	12.3	Masu 2	23.2	11	Gazabure	59.2	24.6
Femari	12.4	14.1	Dagula	20	12.9	Gundi	24.8	8.9
Dikwa 2	19.2	11.5	Hamsuri	6.4	3.3	Kabiya	35.2	16.9

(Continued)

Table 8.5 (Continued).

Giskur	8.4	4.1	Mulige	10	14	Majiri	70	27.7	
Tungushi	15.4	8.1	Gubio	3.7	2.1	Tamsuguwa	53.5	21.9	
Kurnowa	12.4	6.3	Gadai	8.8	5.1	Gashagar	32.5	13.8	
Damakuli	12.8	10.4	Boboshe	15.6	9.1	Shehubultuwa	35	13.7	
Gajigana	12	6.7	Bowata	56	25.5	Bullara	52	12	
Bida	12.8	8.1	Sadawa	23.2	13.1	Ngala	8.8	5.6	
Kinjimerim	11.2	5.1	Kingoa	13.6	7.2	Kala	10.8	4.5	
Mongonu	6.4	2.3	Kwa	10.4	6	Muktu	11.6	6.1	
Gajiram	12.4	5.6	Laraba	63.6	36.5	Ngigo	8.8	3.1	
Sabaswa	7.2	3.7	Garunda	51.2	22.6	Sapte	10.8	3.5	
Logomani	17	9.8	Gudumbali	42.5	19	Fuye	11.2	3.5	
Gajibo	20.8	11.3	Nyau	44.8	19.6	Gamberu	8	4.3	
M. Majia	16.8	8.1	Banowa	46	20.9	Ran	8	4.1	
Kaza	18.8	0.1	Arege	32.2	15.4	Wulgo	7.2	3.6	
Mudu	18.4	10.4	Kanema	76	24.2	Nduffu	10.8	7.7	

Source of data: Barber, (1965).

Table 8.6 Minimum, maximum, mean and standard deviation for Ca and Mg (mg/l) in the Middle Zone Aquifer of the Chad Basin Area.

	N	Minimum	Maximum	Mean	Std. Deviation
Cnica	63	3.70	76.00	24.7683	18.95952
Mg	63	0.10	36.50	11.3873	7.64622
Valid N (listwise)	63	–	–	–	–

Table 8.7 Mean Ca and Mg concentration (mg/l) in the rock aquifer compared with the permissible standards for drinking water in Northern Nigeria and the WHO.

		Rock Hosted Aquifer	Permissible standards of purity for drinking water (Ministry of Health, Northern Nigeria 1958)	Excessive standards of purity for drinking water (Ministry of Health, Northern Nigeria 1958)	WHO minimum permissible limit	WHO maximum permissible limit
Mean Values of Ca	Basement	30.88	75	200	75	200
Mean Values of Mg		8.99	50	150	50	150
Mean Values of Ca	Sedimentary	26.24	75	200	75	200
Mean Values of Mg		13.31	50	150	50	150
Mean Values of Ca	Younger Granites	7.75	75	200	75	200
Mean Values of Mg		3.49	50	150	50	150
Mean Values of Ca	Volcanic	30.18	75	200	75	200
Mean Values of Mg		23.50	50	150	50	150

Table 8.8 Ca and Mg content (mg/l) in the waters of the Younger Granites Aquifer.

Coordinates	Ca	Mg	Coordinates	Ca	Mg	Coordinates	Ca	Mg
09° 49.861N 08° 54.777E	0.91	0.11	09° 46.718N 08° 52.420E	10.66	1.07	09° 45.952N 08° 51.651E	8.04	1.00
09° 49.926N 08° 54.459E	1.16	0.16	09° 47.490N 08° 52.628E	8.86	1.60	09° 52.518N 08° 52.545E	9.52	0.95
09° 49.742N 08° 54.655E	0.54	0.07	09° 52.817N 08° 53.256E	8.64	1.26	09° 53.499N 08° 53.131E	10.20	1.27
09° 46.070N 08° 54.512E	1.41	0.17	09° 52.683N 08° 52.268E	9.38	1.29	09° 53.029N 08° 52.451E	25.60	1.97
09° 47.040N 08° 53.324E	0.62	0.12	09° 53.213N 08° 51.630E	5.99	0.77	09° 54.116N 08° 52.028E	27.82	17.49
09° 46.295N 08° 52.314E	0.69	0.13	09° 53.298N 08° 50.298E	5.06	0.18	09° 53.444N 08° 49.752E	9.10	3.37
09° 47.628N 08° 54.440E	0.68	0.16	09° 52.003N 08° 51.013E	4.17	0.93	09° 51.938N 08° 50.774E	2.28	2.90
09° 47.896N 08° 53.667E	1.80	0.38	09° 51.241N 08° 52.226E	4.73	0.63	09° 51.786N 08° 51.815E	12.02	3.75
09° 47.917N 08° 52.450	7.16	0.75	09° 48.221N 08° 51.709E	9.18	0.33	09° 46.223N 08° 50.704E	26.00	17.90
09° 49.528N 08° 54.629E	0.00	0.00	09° 43.287N 08° 51.603E	6.07	1.25	09° 46.228N 08° 50.660E	28.00	18.04
09° 49.034N 08° 54.180E	0.00	0.00	09° 45.371N 08° 51.586E	5.94	0.79	09° 45.656N 08° 50.928E	9.52	2.67
09° 48.685N 08° 52.980	0.00	0.00	09° 45.534N 08° 51.990E	6.52	0.59	09° 47.327N 08° 51.282E	11.39	1.90
K VOM	16.21	30.82	09° 46.674N 08° 51.727E	8.12	0.27	09° 47.327N 08° 41.374E	12.63	2.35
GANAWURI 1	12.60	25.34	09° 49.706N 08° 53.371E	5.03	0.99	09° 54.116N 08° 52.028E	26.50	17.10
GANAWURI 2	11.68	21.44	09° 50.684N 08° 54.080E	4.02	2.69	09° 51.938N 08° 50.774E	11.50	3.50
GANAWURI 3	13.83	26.84	09° 53.807N 08° 52.956E	4.27	0.17	09° 45.656N 08° 50.928E	9.30	2.00
GANAWURI 4	8.11	19.10	09° 52.518N 08° 52.545E	9.06	2.10	09° 52.877N 08° 53.256E	6.33	2.67
Katako	38.70	2.80	09° 51.513N 08° 49.748E	3.06	2.63	09° 52.683N 08° 51.268E	5.38	1.01
Kufang	6.40	0.90	09° 47.806N 08° 52.003E	3.96	1.87	09° 52.213N 08° 54.080E	5.60	0.75
09° 51.415N 08° 51.925E	5.63	3.40	09° 47.550N 08° 51.660E	3.90	1.87	09° 54.415N 08° 51.925E	3.50	2.89
09° 50.925N 08° 52.172E	7.23	2.78	09° 47.010N 08° 51.508E	3.85	1.90	09° 49.979N 08° 51.479E	3.80	0.60
09° 49.361N 08° 51.877E	10.49	1.24	09° 51.580N 08° 53.900E	3.55	2.70	09° 47.016N 08° 51.716E	4.00	0.55
09° 49.979N 08° 51.479E	5.75	1.92	09° 47.016N 08° 51.716E	2.66	1.82	09° 45.952N 08° 51.651E	3.95	0.30
09° 49.821N 08° 52.119E	5.08	1.34	09° 43.249N 08° 51.979E	5.23	2.56	09° 52.518N 08° 52.545E	4.35	0.20
09° 47.016N 08° 51.716E	6.31	1.25	09° 44.270N 08° 53.134E	5.24	3.83	09° 52.518N 08° 52.543E	4.50	0.20
09° 43.249N 08° 51.514E	6.58	1.15	09° 53.499N 08° 53.134E	9.20	1.20	09° 45.656N 08° 50.928E	1.50	0.30
09° 44.276N 08° 51.514E	4.65	0.89	09° 53.029N 08° 52.451E	20.60	1.50	09° 46.718N 08° 52.422E	4.00	0.30
09° 48.221N 08° 51.700E	4.00	0.25	09° 43.287N 08° 51.823E	2.25	0.10			

Source: Dibal (2012), (2007); Chilota (2011); Chibuzo (2011).

Table 8.9 Minimum, maximum and mean content (mg/l) in the Younger Granites Aquifer.

	N	Minimum	Maximum	Mean	Std. Deviation
Ca	83	0.00	38.70	7.7560	7.15587
Mg	83	0.00	30.82	3.4980	6.60198
Valid N (listwise)	83	–	–	–	–

Table 8.10 Distribution of Ca and Mg (mg/l) in the Volcanic Aquifer.

Longitude	Latitude	Ca	Mg	Longitude	Latitude	Ca	Mg
12.176	10.006	127.30	61.83	12.228	10.52	31.90	14.84
12.158	10.567	56.27	32.72	12.206	10.608	317.60	89.89
12.191	10.592	73.18	42.32	12.207	10.571	62.83	25.33
12.19	10.612	48.76	26.44	12.137	10.59	74.85	37.58
12.134	10.618	33.88	29.52	12.139	10.589	42.03	17.92
12.175	10.625	340.10	198.50	12.14	10.555	46.05	28.47
12.191	10.626	53.06	27.12	12.134	10.542	34.36	18.44
12.19	10.626	67.83	31.80	12.153	10.552	52.54	36.52
12.134	10.575	85.57	52.08	12.169	10.619	0.00	0.00
12.172	10.604	67.77	35.54	12.158	10.573	54.27	24.17
12.157	10.601	75.39	46.54	12.158	10.569	254.80	99.17
12.175	10.6	17.57	7.75	12.201	10.603	0.00	0.00
12.176	10.54	16.92	6.88	12.205	10.586	26.40	9.33
12.174	10.555	32.56	16.56	12.208	10.583	171.00	60.19
12.19	10.609	0.00	0.00	12.205	10.54	282.50	139.90
12.123	10.608	22.27	13.45	12.176	10.54	0.00	0.00
12.124	10.61	299.70	171.30	12.172	10.556	0.00	0.00
12.24	10.526	19.14	7.35	12.24	10.525	332.20	173.50
Ajing	–	14.50	19.70	12.127	10.525	0.00	0.00
Bwonpe	–	14.20	17.50	Kurgwam	–	4.70	2.20
Panyam	–	16.30	16.20	T. Kerang	–	27.60	22.60
Pungkuk	–	22.00	18.30	Kerang North	–	14.30	13.70
Amshal 1	–	14.90	18.20	Amshal 2	–	14.50	16.20
Konji	–	17.90	15.40	Jwakmang	–	24.60	26.40
Ugut	–	21.50	20.50	Pang	–	15.20	21.10

Source of data: Guskit, 2010; A damu, 2012.

Table 8.11 Minimum, maximum and mean values (mg/l) of calcium and magnesium in the Volcanic Hosted Aquifer.

	N	Minimum	Maximum	Mean	Std. Deviation
Ca	50	0	332.20	30.18	22.527
Mg	50	0	173.50	23.50	20.646
Valid N (listwise)	50	–	–	–	–

REFERENCES

Ajibade A.C., Fitches W.R. & Wright J.B. (1979) The Zungeru mylonite. Nigeria: recognition of a major tectonic unit. In: Kogbe C.A. (editor): *Geology of Nigeria*, Rock View, Nigeria limite-Jos, Nigeria, 57–69.

Akolo Y.E. (2011) General Geology and Hydrogeochemistry of Langtang Area. Unpublished BSc Project. Department of Geology and Mining, University of Jos.

Annor A.E. & Freeth S.J. (1985) Thermotectonic evolution of the Basement Complex around Okene, Nigeria, with specific reference to deformation mechanism: *Precambrian Research*, 28, 73–77.

Ashano E.C., Dibal H.U. & Aghomishie, M.A. (2007) Water quality around Jos-Bukuru Metropolis. *Journal of Environmental Sciences* 1(2), 45–53.

Barber W. (1958) Pressure water in the Chad Formation of Bornu and Dikwa Emirate, north east Nigeria, *Geological Survey of Nigeria Bulletin*, 35, 62–68.

Burke K.C. & Dewey J.F. (1972) Orogeny in Africa. In: Dessauage T.F.J., Whiteman A.J. (editors): *African Geology*, University Ibadan Press, 583–608.

Chibuzor E.B. (2011) Geology and water quality of some ponds around Gazon Village and environs, Jos Plateau State, Unpublished BSc Project, Department of Geology and Mining, University of Jos.

Chidinma G.A. (2011) The Geology and hydrogeochemistry of Bapkwai Area. Unpublished BSLT Project, Department of Geology and Mining, University of Jos.

Chilota E.R. (2011) Geology and trace elements content in abandoned mined ponds waters of Bargada Area, Plateau State Unpublished BSc Project, University of Jos, Department of Geology and Mining.

Cyril E.C. (2010) Geology and concentration of trace elements in surface waters of Demakka area. Unpublished BSc Project, Department of Geology and Mining, University of Jos.

Dada S.S., Brick J.L., Lancelot J.R. & Rahaman M.A. (1993) Archaean migmatite-gneiss complex of north central Nigeria: Its geochemistry, petrogenesis and crustal evolution. International colloquium on African Geology, Mbabane, Swaziland, 97–102.

Dada S.S., Briqueu L., Harms U., Lancelot J.R. & Matheis G. (1995) Charnockitic and monzonitic Pan African series from north central Nigeria: Trace element and Nd. Sr, Pb isotope constraints on their petrogenesis: *Chemical Geology*, 124, 233–252.

Dung J.J. (2005) Geology and Nitrate Concentration in the Groundwaters of Dott-Kagadama Area, Bauchi State. Unpublished. BSc Project, Department of Geology and Mining, University of Jos.

Eli J.D. (2005) Geology and fluoride concentration in waters of Bununu-Dass. Unpublished BSc Project, Department of Geology and Mining, University of Jos.

Garba I. (2002) Late Pan African tectonics and origin of gold mineralization and rare meta pegmatites in the Kushaka schist belt, Northwestern Nigeria. *Journal of Mining and Geology*, 38, 1–12.

Guskit R.B. (2010) Major and trace element distribution in natural water waters and soils in Panyam Volcanic Province-Health Implication. Unpublished MSc thesis. Department of Geology and Mining, University of Jos.

Haggai M.U. (2011) Geology, major and trace elements in waters in Langtang: Health Implications. Unpublished BSc Project, University of Jos, Department of Geology and Mining.

Jones H.A. & Hockey R.D. (1964) The geology of parts of south western Nigeria. In: OshinO. (editor): The Basement Complex of Nigeria and its mineral resources. A symposium organized to mark the 60th birthday of Professor M.A.O Rahaman, Ile-Ife, Nigeria.

Kehinde J.M. (2010) Geology and trace elements geochemistry in surface waters of Keana. Unpublished B.Sc Project, Department of Geology and Mining, University of Jos.

Lar U.A., Dibal H.U., Daspan R. & Jaryum S.W. (2007) Fluoride occurrence in the surface and groundwater of Fobur area of Jos East LGA of Plateau States. *Journal of Environmental Sciences* 2, 99–105.

McCurry P. (1971) Pan African Orogeny in Northern Nigeria: *Geological Society Bulletin*, 82, 3251–3263.

McCurry P. (1973) *The Geology of Degree Sheet 21, Zaria, Nigeria*. Overseas Geology and Mineral Resources, London.

McCurry P. (1976) The geology of Precambrian to Lower Paleozoic rocks of Northern Nigeria: In: Kogbe C.A. (editor): *Geology of Nigeria*, Elizabeth Publishing Company, Lagos.

Obinna N.P. (2011) General geology and chemical constituents of waters around Zamzhimin, Zamgbagim and environs. Unpublished BSLT Project, Department of Geology and Mining, University of Jos.

Oche I.R. (2010) Geology. Major and trace elements concentration of Adaga area. Unpublished BSc Project, Department of Geology and Mining, University of Jos.

Russ W. (1957) The Geology of parts of Niger and Zaria, Provinces. *Bulletin Geological. Survey of Nigeria*, 27.

Shola O.G. (2010) The geology and chromium concentration in surface and groundwaters of Ature town. Unpublished BSc Project, Department of Geology and Mining, University of Jos.

Chapter 9

Effect of mine waters from coal mines of the Upper Silesian Coal Basin in the content of Ca and Mg in the catchment of the Upper Odra river

Irena Pluta[1] & Arnost Grmela[2]
[1]KNCh PAN, Katowice, Poland
[2]VSB-Technicka Universita Ostrava, Ostrava-Poruba,
Czech Republic

ABSTRACT

The effects of waters pumped from coal mines of the Upper Silesian Coal Basin in the Czech and Polish parts of the catchment of the Upper Odra on Ca and Mg in waters of the Olza and Odra rivers is presented. Mine waters enrich the water in the Odra river with Mg ion concentration in the range 30 to 125 mg/l within the regulatory requirements of Polish law.

9.1 INTRODUCTION

Waters of various chemical compositions, ranging from fresh waters to brines, flow into the coal mine workings of the in Czech and Polish part of the Upper Silesian Coal Basin (Rozkowski *et al.*, 2004; Labus & Grmela, 2004; Pluta, 2005, 2011). The fresh waters are utilised for potable water, water for industrial processes or can be transferred to surface waters without harm to the aquatic environment. The problem of handling brines, however, is of an enormous scale worldwide (e.g. Lebecka *et al.*, 1986; Pluta, 2005, 2011).

Coal mines in the Upper Silesian Coal Basin are located near the headwaters of the two largest rivers in Poland: the Odra and Vistula. Major pollution of these two rivers is caused by mine waters discharged from coal mines. These waters contain sodium, barium, radium, chlorides and also calcium and magnesium ions. The aim of this chapter is to review the impact of mine waters from coal mines discharged into the catchment of the Upper Odra and Olza catchments on concentrations of Ca and Mg.

9.2 LOCATION OF STUDIED AREAS

The Czech part of the Upper Silesian Coal Basin is located in centre of the northern part of the Czech Republic and the Polish part is located in the southern part of Poland (Figure 9.1). The geological history and hydrogeological conditions of this basin

Figure 9.1 Location of the Upper Silesian Coal Basin.

are well known. The Upper Silesian Coal Basin lies in the Upper Silesian Variscian Intermentone Depression whose geological development resulted from the Variscian and Alpine orogenies (Kotas, 1982; Bula & Kotas, 1994). The molasse sediments of the Upper Carboniferous strata within this depression are about 8200 m thick. In the south western part of the Upper Silesian Coal Basin, where the catchment of the Upper Odra is located, the Carboniferous formation is covered by Miocene and Quaternary deposits. There are more than fourteen deep coal mines are still operating in this region. Natural mine waters of various origin and chemical composition flow into the mine workings. Their origin, determined on the basis of isotope (deuterium, oxygen) and hydrochemical data and isotopic composition of sulphates, shows that the waters formed in the infiltration phases, most likely in the Lower Permian – Rotliengendes, in the Lower Jurassic – Lias and before the marine transgression in Badenian-the Lower Neogene-Paleogene or Early Miocene (Pluta & Zuber, 1995; Pluta, 2005). Reconstruction of the analysis of morphological discordances of different age in the uppermost Carboniferous formations and climate conditions can be used to support this conclusion. Mine waters contain many chemical components mainly Ca and Mg. The concentration of the Ca ion is up to 12 360 mg/l and Mg ion up to 11 540 mg/l (Pluta & Cebulak, 2010).

9.3   MATERIALS AND METHODS

Mine waters are discharged into the Olza and Odra rivers from the Czech coal mines Paskov, Karwina, CSM and Darkov. Mine waters from Polish coal mines in the catchment of the Upper Odra are discharged into the Odra river in Krzyzanowice by a hydro-technical system. The system was built to protect the Odra river waters against con-taminants flowing from the mine waters. It consists of a pipeline of about 80 km length and several retention-feeder installations. Mine waters from eight coal mines, Borynia, Chwalowice, Jankowice, Jastrzebie, Krupinski, Marcel, Pniowek and Zofiowka, flow through the pipeline in controlled system into the Odra river in Krzyzanowice.

Samples of waters were taken in Olza (Godow) and Odra (Chalupki, Krzyzanowice and Raciborz) (Figure 9.2). Concentration of the Ca and Mg ions in were measured three times from November 2010 to June 2011.

The concentration of Ca^{2+} and Mg^{2+} was determined by atomic absorption spectrophotometry ASA in accordance with Polish norm: PN-EN ISO 7980 at the Central Mining Institute.

9.3.1 Polish regulations on Ca and Mg ions in waters

According to the regulation in Polish law, announcements of 29 March 2007 by the Minister of Health (Dz.U. 2007, No 61, 417) on the scope of requirements needed for the quality of drinking water, recommends not more than 30 mg/l of Mg if the concen-tration of sulphates (VI) >250 mg/l and concentration 125 mg/l under conditions where the sulphate (VI) concentration is lower. A concentration of calcium carbonate from 60 to 600 mg/l (calcium from 24 to 240 mg/l) is permitted. These recommendations con-firm medicinal properties and are not obligatory for sewage and water discharges.

The threshold limit values for Ca and Mg in surface waters first and second cleanliness class is set out in the regulations announcement 9 November 2011 by the Minister of Environment (Dz.U. 2011, No 257, 1545). The values for Ca are 100 and 200 mg/l and magnesium 50 and 100 mg/l.

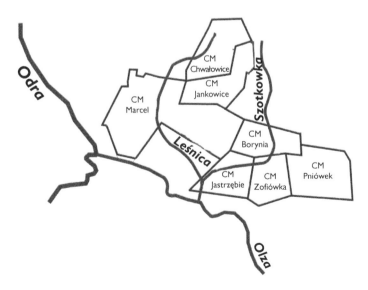

Figure 9.2 Location of the studied area.

9.4   RESULTS

Waters in Olza river from Godow and in Odra river from Chalupki, Krzyzanowice and Raciborz were sampled in November 2010, March and June 2011. Concentrations of Ca and Mg ions in these waters are presented in Tables 9.1a, b and 9.2a, b.

Average values of concentrations of Ca and Mg ions in waters of Olza in Godow and of the Odra in Chałupki, Krzyzanowice and Raciborz are illustrated in Figure 9.3.

It was found that the amount of ions introduced to the waters of the Odra river ranges:

- in Chałupki from 40 to 66 mg-Ca/l and from 10 to 18 mg-Mg/l
- in Krzyzanowice from 54 to 90 mg-Ca/l and from 15 to 30 mg-Mg/l
- in Raciborz from 40 to 66 mg-Ca/l and from 15 to 30 mg-Mg/l.

Table 9.1a Concentrations of Ca^{2+} (mg/l) in waters from Olza river.

Sampling site	Sampling date		
	11.2010	*03.2011*	*06.2011*
Godów	59.5	49.5	69.4

Table 9.1b Concentrations of Mg^{2+} (mg/l) in waters from Olza river.

Sampling site	Sampling date		
	11.2010	*03.2011*	*06.2011*
Godów	12.5	8.8	14.6

Table 9.2a Concentrations of Ca^{2+} (mg/l) in waters from Odra river.

Sampling site	Sampling date		
	11.2010	*03.2011*	*06.2011*
Chałupki	55.5	40.4	65.6
Krzyżanowice	71.2	53.8	89.8
Racibórz	74.5	55.6	88.4

Table 9.2b Concentrations of Mg^{2+} (mg/l) in waters from Odra river.

Sampling site	Sampling date		
	11.2010	*03.2011*	*06.2011*
Chałupki	13.4	9.5	18.2
Krzyżanowice	20.3	15.3	30.1
Racibórz	20.1	14.8	30.2

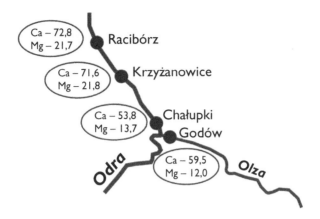

Figure 9.3 Average concentrations of Ca and Mg ions (mg/l) in waters of the Olza and Odra rivers.

Amounts of Ca and Mg in the Olza river (Godow) and the Odra river (Chalupki) were effected by mine waters pumped from Czech coal mines.

Mine water introduced to the Odra river in Krzyzanowice was from eight Polish coal mines and was between 13 to 26 mg-Ca/l, and from 5 to 16 mg-Mg/l.

In the Olza river waters the concentration of Ca is almost five times higher than the concentration of Mg, but in the Odra river waters the ratio of concentrations is from 3.5 to 4.

Concentrations of Ca and Mg ions in the water of the Olza at Godow and in Odra at Chałupki are almost the same as in the Odra river in Krzyznowice and Raciborz.

9.5   CONCLUSIONS

Taking into consideration that water from the Odra river should meet the standards of surface waters used for agriculture and domestic needs the systematic and through investigations on the Ca and Mg ions were performed between November 2010 and June 2011. It was found that the concentration of ions introduced into the Odra river at Krzyzanowice was in the range 13 to 26 mg-Ca/l, and from 5 to 16 mg-Mg/l.

In general, mine waters from coal mines of the Upper Silesian Coal Basin, discharged in the catchment of the Upper Odra enrich the waters of the river by a concentration range from 30 to 125 mg-Mg/l but within the needs of Polish law.

REFERENCES

Buła Z. & Kotas A. (1994) *Geological atlas of the Upper Silesian Coal Basin, part III, Geological-structure maps.* Ed. PIG Warszawa.

Dz.U. No 61, item 417 (2007) *Order the Minister of Health on quality drinking water requirements.*

Dz.U. No 257 item 1545 (2011) *Order the Minister of Environmental on mode of classification of surface waters and environmental norm quality for priority substances.*

Kotas A. (1982) *Sketch of the geological structure of the Upper Silesian Coal Basin* (in Polish). In: Przewodnik 54 Zjazdu Polskiego Towarzystwa Geologicznego. Ed. Wyd. Geologiczne Sosnowiec-Warszawa.

Labus K. & Grmela A. (2004) Isotopic composition of ground water in the SW part of the Upper Silesian Coal Basin within territories of Poland and Czech Republic. *Sbornik vedeckych praci Vysoke skoly banske-Technickie univerzity Ostrava.* C1, 57–68.

Pluta I. (2005) Waters of the Upper Silesian Coal Basin- origin, pollutants and purification (in Polish), *Prace Naukowe GIG*, 865, 169.

Pluta I. (2011) *Hydrogeochemistry of Carboniferous formation in south part of the Upper Silesian Coal Basin* (in Polish). Ed. GIG Katowice.

Pluta I. & Cebulak S. (2010) Oxyreactivity of coals and environmental of waters in selected parts of the Marcel Coal Mine (in Polish). *Przegląd Górniczy* 1–2, 101–105.

Pluta I. & Zuber A. (1995) Origin of brines in the Upper Silesian Coal Basin (Poland) inferred from stable isotope and chemical data. *Applied Geochemistry* 10, 447–460.

Różkowski A., *et al.* (2004) *Hydrogeochemical environment in Carboniferous formations of the Upper Silesian Coal Basin* (in Polish), Ed. USl, Katowice.

Part 2

Significance of calcium and magnesium in groundwater for human health

Chapter 10

Health and regulatory aspects of calcium and magnesium in drinking water

Frantisek Kozisek

Department of Water Hygiene, National Institute of Public Health, Prague, Czech Republic

ABSTRACT

The chapter reviews the history of understanding the health significance of Ca and Mg in water over the last 150 years. The main focus is to provide an updated review of the scientific evidence of health effects from either low or high concentrations of Ca and Mg in drinking water and the health and aesthetic benefits to be gained from 'optimum concentrations' of both elements. Problems with insufficient regulation of both elements in drinking water are discussed and possible measures to ensure the minimum and optimum Ca and Mg levels in drinking water are recommended.

10.1 INTRODUCTION

Calcium (Ca) and magnesium (Mg) as major constituents of natural water have been extensively studied, but surprisingly, the extensive data on the beneficial health effects have had a low impact in the regulatory context. This is not satisfactory from a public health point of view, because neglecting this fact facilitates distribution of drinking water without an optimum mineral balance and contributes to deterioration of health of individuals or populations regularly exposed to such water. Both Ca and Mg are essential elements to secure the vital functions of the human body, but, based on the survey data available, it is clear that very large numbers of people consume levels of these minerals that are insufficient to support even the most conservative estimates of their physiological needs (WHO, 2009).

Ca and Mg in drinking water relates also to a third parameter, that of water hardness, even if this term is incorrect and obsolete from a strictly chemical point of view. Both of these elements largely have not been analysed individually in drinking water in the past, but rather non-specifically in summary form as hardness. This approach was applied to many studies focused on health effects. Initially, water hardness was understood to be a measure of the capacity of water to precipitate soap, which is in practice the sum of concentrations of all polyvalent cations present in water (including Ca, Mg, Sr, Ba, Fe, Al and Mn). Since the other ions (apart from Ca and Mg) play a minor role in this regard, it has been generally accepted that hardness is defined as the sum of the Ca and Mg concentrations, determined by the EDTA titrimetric method, and expressed in mmol/l (ISO, 1984) or as $CaCO_3$ equivalent in mg/l (APHA/AWWA/WEF, 1998), and less frequently as the CaO equivalent.

It is generally acknowledged that research on the health effects of water hardness over the last 50 years started with the paper by the Japanese chemist Kobayashi, who showed, based on epidemiological analysis, higher mortality rates from stroke in the areas of Japanese rivers with more acid water compared to those with more alkaline, i.e. harder water used for drinking purposes (Kobayashi, 1957). Interest is in fact much older and may be traced back before World War I (Thresh, 1913) and even to the 1870s when Dr Henry Letheby, Medical Officer of Health for the City of London, studied the relationships between total mortality and water hardness in 19 cities in England and Scotland (Anonymous, 1871).

Results of number of epidemiologic studies carried out in the 1960s–1970s were summarised in the compelling dictum 'soft water, hard arteries' (i.e. if the water is low in calcium and magnesium, higher incidence of cardiovascular diseases was observed in the population supplied). This was widely accepted by both water and public health experts and many health agencies, including the World Health Organization (WHO), in the 1980s. Although this knowledge was later called into question by some, more recent scientific literature has not rejected this dictum, but brought new evidence as well additional information on other possible beneficial health outcomes beside the cardiovascular diseases.

10.2   HEALTH EFFECTS OF CALCIUM AND MAGNESIUM IN DRINKING WATER

Over 99% of total body Ca is found in bones and teeth, where it functions as a key structural element. The remaining body Ca functions in metabolism, serving as a signal for vital physiological processes, including vascular contraction, blood clotting, muscle contraction and nerve transmission. Inadequate intakes of Ca have been associated with increased risks of osteoporosis, nephrolithiasis (kidney stones), colorectal cancer, hypertension and stroke, coronary artery disease, insulin resistance and obesity. Ca is unique among nutrients, in that the body's reserve is also functional: increasing bone mass is linearly related to reduction in fracture risk. The recommended Ca daily intake for adults ranges between 1000 and 1200 mg. Some population groups may need a higher intake (WHO, 2009).

Mg is a cofactor for some 350 cellular enzymes, many of which are involved in energy metabolism. It is also involved in protein and nucleic acid synthesis and is needed for normal vascular tone and insulin sensitivity. Low magnesium levels are associated with endothelial dysfunction, increased vascular reactions, elevated circulating levels of C-reactive protein and decreased insulin sensitivity. Low Mg status has been implicated in hypertension, coronary heart disease, type 2 diabetes mellitus and metabolic syndrome. The recommended magnesium daily intake for an adult is about 250–350 mg (WHO, 2009).

The same principles apply to Ca and Mg in drinking water as for essential elements intake generally: either very low or very high intake (concentration in water) may represent some risk for human health, while 'medium' or optimum intake or water concentration is relating to beneficial effects and support both human health and aesthetic acceptability of such water.

Effects of drinking water hardness were most frequently studied because of cardiovascular diseases. The most comprehensive review of these studies was commissioned by the Drinking Water Inspectorate of England and Wales and completed by the University of East Anglia in 2005 (Catling *et al.*, 2005). The initial search identified 2096 papers. Papers were retained if they presented primary data of human studies, were directly related to the research question (to review and critically assess the merits of available studies concerning the health effects of soft and softened water on cardiovascular disease and cancer) and involved a comparison of populations or individuals at different levels of exposure. Experimental animal studies and human dietary studies were excluded from this review. Overall, 132 studies were identified as primary data papers, of which 17 papers were excluded due to their descriptive but non-analytical content. A total of 115 underwent full article appraisal by two independent reviewers.

The majority of studies reviewed were of an ecological study design. Further study quality criteria were applied to categorise the ecological studies by high, medium or low quality. A total of 60 such papers were evaluated, of which 44 met the minimum quality criteria. Of the 12 high quality studies, 9 presented evidence for a significant inverse association between water hardness, Ca and/or Mg levels and cardiovascular mortality. The remaining three studies found no significant association. Of the 32 medium and low quality studies, 22 found a significant inverse association. Ecological studies are currently considered to provide only limited evidence as individual exposure is not assessed, but they bring additional supporting information to more advanced studies.

The more advanced epidemiological studies (cross-sectional studies, case control studies, and cohort studies), showed the following results. Five cross sectional studies were identified, of which only 2 sampled the drinking water quality at the individual level with the remainder using an ecological measure of the drinking water parameters. These papers examined individual level cardiovascular risk factors with an inverse association between drinking water Ca and/or Mg and blood pressure and serum lipids observed in some, but not all studies. Six case control studies examined both drinking water Mg and Ca and risk of death from cardiovascular disease. Of these, four found a significant inverse association with Mg concentrations. Of three cohort studies reviewed, two were of medium/poor quality and used an ecological measure of drinking water factors and limited or non-existent control for possible confounders. The third study was conducted in Great Britain and found no association between drinking water hardness and cardiovascular disease. However this study also suffered from poor exposure characterisation.

Subsequent systematic review and meta-analysis of 14 analytical observational studies (i.e. the most valid epidemiological studies) investigating the association between cardiovascular disease and drinking water hardness brought convincing epidemiological evidence about the protective role of Mg in drinking water as a pooled odds ratio and showed a statistically significant inverse association between Mg and cardiovascular mortality (OR 0.75 (95% CI 0.68, 0.82), p, 0.001). It means that the highest exposure category (people consuming drinking water with magnesium 8.3–19.4 mg/l) was significantly associated with a decreased likelihood of cardiovascular mortality (by 25%), compared with the baseline, i.e. people drinking water with

Mg content of 2.5–8.2 mg/l (Catling *et al.*, 2008). The protective role of water Ca towards cardiovascular disease was also confirmed by some studies, but the evidence is not as strong as for magnesium (Catling *et al.*, 2008).

A number of other papers suggest a beneficial or protective effect of water Ca and Mg on other diseases. These include Ca in neurological disturbances (Jacqmin *et al.*, 1994; Emsley *et al.*, 2000), Ca and Mg in amyotrophic lateral sclerosis (Yasui *et al.*, 1997), Mg in preeclampsia in pregnant women (Melles & Kiss, 1992), Ca in high blood pressure (Rubenowitz *et al.*, 1999), Mg in high blood pressure and metabolic syndrome (Rasic-Milutinovic *et al.*, 2012). Costi *et al.* (1999) concluded that a regular life-long daily intake of drinking water with highly bio-available Ca may be of importance for maintaining the calcium balance and improving the spinal bone mass. Calcium mineral water supplementation for one year showed an increase in the bone mass density in postmenopausal women (Cepollaro *et al.*, 1996). An association between low Ca content of drinking water and higher incidence of fractures in children was found in Spain (Verd Vallespir *et al.*, 1992).

Since the late 1990s a number of epidemiological studies (mostly combined ecologic case-control studies) were carried out in Taiwan by one research team to focus on relationships between drinking water hardness and mortality from various types of cancer. It showed a significant geographical variation. Mg was found to have a protective effect against cerebrovascular diseases (Yang, 1998) and hypertension (Yang & Chiu, 1999b), water hardness showed a protective effect against cardiovascular disease (Yang *et al.*, 1996), cancer of oesophagus (Yang *et al.*, 1999c), cancer of pancreas (Yang *et al.*, 1999d), cancer of rectum (Yang *et al.*, 1999e) and breast cancer (Yang *et al.*, 2000). Drinking water Ca proved protective against colorectal cancer (Yang *et al.*, 1997) and gastric cancer (Yang *et al.*, 1998). The risk of rectal cancer from trihalomethanes (THM) was increased when the Mg level was low in drinking water (Kuo *et al.*, 2010); and similarly the risk of kidney cancer from THM was higher in soft water areas (Liao *et al.*, 2012). If the Mg concentration was higher than around 17 mg/L there were fewer deaths due to diabetes (Yang *et al.*, 1999a). However, further studies from other countries are needed to confirm these results. The protective effect of Mg in drinking water on decreased incidence of Type 2 diabetes among young adults was confirmed by one study in Finland (Kousa *et al.*, 2012).

Ca and to a lower extent also Mg in both drinking water and food were previously found to have a beneficial antitoxic effect since they prevent – via either a direct reaction resulting in an nonabsorbable compound or competition for binding sites – absorption or reduce harmful effects of some toxic elements such as lead and cadmium (Kozisek & Rosborg, 2008).

There is no evidence about any harmful health effects if Ca is present in drinking water below 200 mg/l and Mg below 100 mg/l. Perhaps only a high Mg content (hundreds of mg/l) coupled with a high SO_4 content (above 500 mg/l) may cause transient diarrhoea (WHO, 2011a). Nevertheless, such cases are rare. Other harmful health effects (e.g. higher incidence rates of cholelithiasis, urolithiasis, arthrosis and arthropathies) caused by hard water are questionable. These effects were observed in some older Russian ecological epidemiological studies (Muzalevskaya *et al.*, 1993; Golubev & Zimin, 1994; Mudryi, 1999) due to high water hardness, i.e. more than 5 mmol/l, but these waters were at the same time rich in total dissolved solids (above 1000 mg/l) showing mineral levels which are not typical of most drinking waters.

It is not possible to determine if these effects are caused solely by Ca and Mg (probably not) or by other cation(s) and anion(s) present or by total dissolved solids. The literature describes an exceptional case of urolithiasis in an infant in France to whom food was prepared only from bottled natural mineral water with high content of Ca (555 mg/l) and Mg (110 mg/l) (Saulnier *et al.*, 2000).

Hard water is also rarely reported to cause an increase in the risk of atopic eczema in school children (McNally *et al.*, 1998; Miyake *et al.*, 2004; Chaumont *et al.*, 2012). This can probably be explained by its higher drying effect on the skin (similar to that of over-chlorinated water) or higher soap consumption, but in this case water is used externally and not for drinking.

Higher water hardness may worsen aesthetic (organoleptic) characteristics of drinking water or drinks and meals prepared with such water. It can form a layer on the surface of coffee or tea, cause a loss of aromatic substances from meals and drinks (due to bonding to calcium carbonate), create an unpleasant taste in the water itself for some consumers (calcium taste threshold is about 100–300 mg/l, unpleasant taste starts from 500 mg/l). Taste also depends upon the presence of other ions; the magnesium content exceeding 170 mg/l together with the presence of chloride and sulphate anions are responsible for the bitter taste of water (WHO, 2011a). Very soft water, such as distilled, osmotic, and rain water as two extreme examples, is of unacceptable taste for most people who usually report them to be an unpleasant soapy taste. It means that a certain minimum concentrations of minerals, the most crucial of which are Ca and Mg salts, are essential for the pleasant and refreshing taste of drinking water and very high or low mineral content may be considered unpalatable by consumers (WHO, 2009).

Based on the available data, the desirable minimum concentrations of Mg and Ca in drinking water can be estimated to be about 10 mg/l (for Mg) and about 20–30 mg/l (for Ca), respectively. Nevertheless, this does not mean that if low levels of these elements were increased to remain below these minima (e.g. if the magnesium level were increased from 2 to 5 mg/l), it would be of no importance. It seems that any increase of Ca or Mg in soft water, even by several mg/l, could have a beneficial or protective health effect (in addition to the known technical effect – reduction of aggressiveness of the water). Although a certain minimum quantity of these elements is desirable, it definitely does not mean the more the better. What can be called the optimum levels in drinking water (from a health point of view) ranges from 20 to 30 mg/l for magnesium and from 40 to 80 mg/l for calcium, respectively, and for water hardness as Σ Ca + Mg from about 2 to 4 mmol/l (Kozisek, 2005).

As Ca and Mg are metabolic antagonists, not only is the absolute content of both elements in water (diet) is important, but so also is the Ca:Mg ratio. To allow the best absorption rate of both elements, the Ca:Mg intake ratio (and drinking water ratio) should ideally be 2–3 (2/1 to 3/1) (Durlach *et al.*, 1989).

10.3   REGULATORY ASPECTS OF CALCIUM AND MAGNESIUM IN DRINKING WATER

Considering the high number of epidemiological studies confirming the beneficial effects of certain amounts of Ca and Mg in drinking water and the large body of supporting evidence from experimental and clinical studies, as well as zero health risk relating to usual levels found in drinking water (dozens of mg/l), it is surprising to see the restrained attitude of the World Health Organization (WHO) over the last 20 years to recommend any guideline value. It is surprising if we realise that in 1970s and 1980s the WHO acknowledged the importance of water hardness for population health. One can just try to guess the true motives behind the current WHO position. It is even more surprising to read background documents (Hardness in Drinking-water) for development of WHO *Guidelines for Drinking-water Quality* from 1996 (WHO, 1996) to 2011 (WHO, 2011b) and to find that methodically poor epidemiological studies are referred to support importance of water hardness, while advanced studies are mentioned only as footnote or omitted (there are also supporting animal experimental studies or clinical studies which are also not taken into account), and to read conclusion: '*Although there is some evidence from epidemiological studies for a protective effect of magnesium or hardness on cardiovascular mortality, the evidence is being debated and does not prove causality. Further studies are being conducted. There are insufficient data to suggest either minimum or maximum concentrations of minerals at this time, and so no guideline values are proposed.*' (WHO, 2011b).

This view is in contrast to the statement of two prominent epidemiologists R. Calderon and P. Hunter who concluded their chapter on epidemiological studies and the association of cardiovascular disease risks with water hardness in the WHO monograph on Ca and Mg in drinking water: '*Information from toxicological, dietary and epidemiological studies supports the hypothesis that a low intake of magnesium may increase the risk of dying from, and possibly developing, cardiovascular disease or stroke. Thus, not removing magnesium from drinking-water, or in certain situations increasing the magnesium intake from water, may be beneficial, especially for populations with an insufficient dietary intake of the mineral. This raises a significant policy issue. How strong does the epidemiological and other evidence need to be before society acts to reduce a potential public health threat rather than await further evidence that such a threat is real? Such a decision is a political rather than a purely public health issue. There is a growing consensus among epidemiologists that the epidemiological evidence, along with clinical and nutritional evidence, is already strong enough to suggest that new guidance should be issued.*' (Calderon & Hunter, 2009).

It seems that a deficiency of Ca and Mg in drinking water poses at least a comparable health risk to exceeding the limit for some toxic substances (which are regulated, even though the evidence of their toxicity is much less convincing than evidence of beneficial effect of Ca or Mg). Nevertheless, the precautionary principle is not applied by the WHO in the case of Ca and Mg, or at least Mg, where the evidence is much stronger.

Introduction of regulatory measures concerning the minimum levels of Ca and Mg in drinking water seems to be justified and highly desirable. They should be based on the fact that it is much simpler and much more effective to keep the existing Ca and Mg drinking water levels than to add these minerals to water artificially. Practically, this means restricting the use of technologies leading to the removal of Ca and Mg from water only to the cases where the Mg and Ca levels are too high (i.e. of hundreds of mg/l or more) provided that the required minimum of Σ Ca + Mg is retained in the water after treatment. A certain requirement for the minimum required concentration of hardness (≥60 mg/l as calcium or equivalent cations) for softened and desalinated water, set up in Council Directive 80/778/EEC (EC, 1980) appeared obligatorily in national legislation of all EEC members in the past. Nevertheless, this Directive was in force only to December 2003, as Directive 98/83/EC replaced it since 1998 (EU, 1998). The latter directive does not present any requirement for the Ca and Mg levels or water hardness (apart from the lower limit for pH ≥ 6.5 which requires indirectly a certain level of dissolved solids); on the other hand, it does not prevent the member states from implementing such a requirement, if needed, into their national legislation.

Nevertheless, apart from the WHO approach and EU Drinking Water Directive (98/93/EC), more than 10 European countries have established some form of minimum requirements on hardness level after softening or a generally optimum range (e.g. Austria, Belgium, Czech Republic, Denmark, Germany, Hungary, Italy, Netherlands, Poland, Slovakia, Sweden, Switzerland). Some countries have these requirements legally based, while others issued recommendations in form of technical standards or guidelines. Other countries try to educate the consumers through information leaflets or websites how to use any softening device with respect to keeping Ca and Mg in the water for drinking and cooking purposes (UK).

10.4 CONCLUSIONS

Ca and Mg are important components of drinking water and are of both direct and indirect health significance. A certain minimum amount of these elements in drinking water is desirable for technical (decreased corrosivity of water), health and aesthetic reasons. Minimum and optimum levels of Ca and Mg in drinking water were suggested above.

There are several options how to ensure the minimum and optimum Ca and Mg levels in drinking water:

a   To select an adequate water source. If several water sources are available or can be blended, preference should be given to the sources (as a rule, to the underground sources) containing the optimum, or at least the minimum, Mg and Ca levels, as considered in the context of general water composition. These sources should be exploited for the drinking (nutritional) purpose rather than for other, technical purposes.

b   To set strict rules for water treatment technologies decreasing the amount of Ca or Mg in drinking water (e.g. distillation, membrane technologies such as reverse osmosis, ion exchange, precipitation, etc.) or to keep some minimum content of Mg and Ca in case of water softening or desalination. To soften drinking water only if needed for health reasons, i.e. not for technical reasons. In the light of rapid growth in membrane technologies and their applicability to drinking water treatment, such rules will be more and more urgently needed to avoid considerable health problems.

c   To promote stabilisation of soft water sources. This procedure is often used to reduce water corrosivity either by passing the water through a $CaCO_3$ filter (sometimes preceded by dosing with CO_2) or by adding a Ca compound such as lime milk directly to water. Unfortunately, this results only in a negligible increase in the Mg level. Nevertheless, this procedure can be at least partly optimised, on the one hand, by means of improvement of the treatment design, and on the other hand, by selection of an adequate filtration material with a higher magnesium content, e.g. material on a basis of $CaCO_3 + MgCO_3$ or $CaCO_3 + MgO$.

d   The issue of increasing the Mg level by the addition of Mg salts directly to water while treated which has not been much tested in practice, but recent case studies from some countries (Israel, Czech Republic) have shown that central fortification with Mg is technically feasible. This step is necessary if water is initially completely desalinated when producing drinking water from seawater.

Public health education seems to be an easily applicable measure for the moment. The public, in particular those living in the areas supplied with water low in Ca or Mg, should be discouraged from using water softeners or other home water treatment units removing Ca or Mg from water intended for drinking and cooking. At the same time, the consumption of Ca/Mg-rich water (e.g. some bottled natural mineral water) could be encouraged to replace at least partly tap (well) water low in the minerals. However, this solution is not easily applicable to water for cooking.

Introduction of regulatory measures concerning the minimum levels of Ca and Mg in drinking water seems to be justified and highly desirable. They should be based on the fact that it is much simpler and much more effective to keep the existing Ca and Mg drinking water levels than to add these minerals to water artificially. Practically, this means restricting the use of technologies leading to removal of Ca and Mg from water only to the cases where the Mg and Ca levels are too high (i.e. of hundreds of mg/l or more) provided that the required minimum of calcium (20–30 mg/l) and magnesium (10 mg/l) is kept in the water after treatment. Nevertheless, apart from this 'negative' regulation, a positive approach should be adopted. If Directive 98/83/EC contains some general instructions concerning e.g. the disinfection by-products: *'where possible, without compromising disinfection, Member States should strive for a lower value,'* why could not it give a general instruction that drinking water should contain certain minimum Mg and Ca levels and that the member states should strive to reach these levels?

Further studies are needed to address not only the traditional issues such as the Ca and Mg levels but also water treatment technologies (e.g. magnetic treatment or phosphate dosing) which do not modify the absolute levels of these elements in water but may mask the presence or may limit the bioavailability and thus beneficial effect of these elements through different mechanisms.

REFERENCES

Anonymous (1971) Der Wasserconsum Londons im Jahre 1869/70 (Water consumption in London in 1869/70). *Journal für Gasbeleuchtung und Wasserversorgung* 14(11), 403–405.

APHA/AWWA/WEF (1998) *Standard Methods for the Examination of Water and Wastewater #2340 Hardness*. 20th ed. American Public Health Association/American Water Works Association/Water Environment Federation, Washington, DC.

Calderon R. & Hunter P. (2009) Epidemiological studies and the association of cardiovascular dinase risks with water hardness. In: *Calcium and Magnesium in Drinking-water*; 110–144. World Health Organization, Geneva.

Catling L.A., Abubakar I., Lake I.R., Swift L. & Hunter P. (2005) *Review of Evidence for Relationship between Incidence of Cardiovascular Disease and Water Hardness*. Drinking Water Inspectorate, London, 142p.

Catling L.A., Abubakar I., Lake I.R., Swift L. & Hunter P.R. (2008) A systematic review of analytical observational studies investigating the association between cardiovascular disease and drinking water hardness. *Journal of Water and Health* 6, 433–442.

Cepollaro C., Orlandi G., Gonnelli S., Ferrucci G., Arditti J.C., Borracelli D., Toti E. & Gennari C. (1996) Effect of calcium supplementation as a high-calcium mineral water on bone loss in early postmenopausal women. *Calcified Tissue International* 59, 238–239.

Chaumont A., Voisin C., Sardella A. & Bernard A. (2012) Interactions between domestic water hardness, infant swimming and atopy in the development of childhood eczema. *Environmental Research* 116, 52–57.

Costi D., Calcaterra P.G., Iori N., Vourna S., Nappi G. & Passeri M. (1999) Importance of bioavailable calcium in drinking water for the maintenance of bone mass in post-menopausal women. *Journal of Endocrinological Investigation* 22, 852–856.

Durlach J., Bara M. & Guiet-Bara A. (1989) Magnesium level in drinking water: its importance in cardiovascular risk. In: Itokawa Y., Durlach J. (eds). *Magnesium in Health and Disease*. J. Libbey & Co Ltd, London; 173–182.

EC (1980) Council Directive 80/778/EEC of 15 July 1980 relating to the quality of water intended for human consumption. *Official Journal of the European Communities* L 229, 30.8.1980, 11–29.

Emsley C.L., Gao S., Li Y., Liang C., Ji R., Hall K.S., Cao J., Ma F., Wu Y., Ying P., Zhang Y., Sun S., Unverzagt F.W., Slemenda C.W. & Hendrie H.C. (2000) Trace element levels in drinking water and cognitive function among elderly Chinese. *American Journal of Epidemiology* 151, 913–920.

EU (1998) Council Directive 98/83/EC of 3 November 1998 on the quality of water intended for human consumption. *Official Journal of the European Communities* L 330, 5.12.1998, 32–54.

Golubev I.M. & Zimin V.P. (1994) On the standard of total hardness in drinking water (in Russian). *Gigiena I Sanitariia* No. 3/1994, 22–23.

ISO (1984) *International Standard ISO 6059. Water quality – Determination of the sum of calcium and magnesium – EDTA titrimetric method*. International Organization for Standardization, Geneva.

Jacqmin H., Commenges D., Letenneur L., Barberger-Gateau P. & Dartigues J.F. (1994) Components of drinking water and risk of cognitive impairment in the elderly. *American Journal of Epidemiology* 139, 48–57.

Kobayashi J. (1957) On geographical relationship between the chemical nature of river water and death rate from apoplexy. *Berichte des Ohara Instituts für landwirtschaftliche Biologie Okyama University* 11, 12–21.

Kousa A., Puustinen N., Karvonen M. & Moltchanova E. (2012) The regional association of rising type 2 diabetes incidence with magnesium in drinking water among young adults. *Environmental Research* 112, 126–128.

Kozisek F. (2005) Health risks from drinking demineralised water. In: *Nutrients in Drinking Water*; 148–163. World Health Organization, Geneva.

Kozisek F. & Rosborg I. (2008) Water hardness may reduce the toxicity of metals in drinking water. In: International Conference *METEAU – Metals and Related Substance in Drinking Water*, Antalya, 24–26 October 2007; Proceedings Book; Cost Action 637. Brussels, 224–226.

Kuo H.W., Chen P.S., Ho S.C., Wang L.Y. & Yang C.Y. (2010) Trihalomethanes in drinking water and the risk of death from rectal cancer: does hardness in drinking water matter? *Journal of Toxicology and environmental Health, Part A* 73(12), 807–818.

Liao Y.H., Chen Ch.Ch., Chang Ch.Ch., Peng Ch.Y. & Chiu H.F. *et al.* (2012) Trihalomethanes in drinking water and the risk of death from kidney cancer: does hardness in drinking water matter? *Journal of Toxicology and Environmental Health, Part A* 75, 340–350.

McNally N.J., Williams H.C., Phillips D.R., Smallman-Raynor M., Lewis S., Venn A. & Britton J. (1998) Atopic eczema and domestic water hardness. *Lancet* 352, 527–531.

Melles Z. & Kiss S.A. (1992) Influence of the magnesium content of drinking water and of magnesium therapy on the occurrence of preeclampsia. *Magnesium Research* 5, 277–279.

Miyake Y., Yokoyama T., Yura A., Iki M. & Shimizu T. (2004) Ecological association of water hardness with prevalence of childhood atopic dermatitis in a Japanese urban area. *Environmental Research* 94, 33–37.

Mudryi I.V. (1999) Effects of the mineral composition of drinking water on the population's health (review) (in Russian). *Gigiena I Sanitariia* No.1/1999: 15–18.

Muzalevskaya L.S., Lobkovskii A.G. & Kukarina N.I. (1993) Incidence of chole- and neph-rolithiasis, osteoarthrosis, and salt arthropathies and drinking water hardness (in Russian). *Gigiena I Sanitariia* No. 12/1993: 17–20.

Rasic-Milutinovic Z., Perunicic-Pekovic G., Jovanovic D., Gluvic Z. & Cankovic-Kadijevic M. (2012) Association of blood pressure and metabolic syndrome components with magnesium levels in drinking water in some Serbian municipalities. *Journal of Water and Health* 10, 161–169.

Rubenowitz E., Axelsson G. & Rylander R. (1999) Magnesium and calcium in drinking water and death from acute myocardial infarction in women. *Epidemiology* 10, 31–36.

Saulnier J.P., Podevin G., Berthier M., Levard G. & Oriot D. (2000) Calcul coralliforme du nourrisson lié à la prise exclusive d'eau minérale riche en calcium (Staghorn lithiasis in an infant related to mineral water high in calcium). *Archives de Pediatrie* 7, 1300–1303.

Thresh J.C. (1913) Hard v. soft water. *The Lancet* 182(4702), 1057–1058.

Verd Vallespir S., Domingues Sanches J., Gonzales Quintial M., Vidal Mas M., Mariano Soler A.C., de Roque Company C. & Sevilla Marcos J.M. (1992) Association between calcium content of drinking water and fractures in children. (In Spanish.) *Anales Espanoles de Pediatria* 37, 461–465.

WHO (1996) *Guidelines for Drinking-water Quality.* 2nd Ed. Volume 2 – *Health Criteria and Other Supporting Information.* World Health Organization, Geneva.

WHO (2009) *Calcium and Magnesium in Drinking-water.* World Health Organization, Geneva.

WHO (2011a) *Guidelines for Drinking-water Quality.* 4th. ed. World Health Organization, Geneva.

WHO (2011b) *Hardness in Drinking-water. Background document for development of WHO Guidelines for Drinking-water Quality.* WHO/HSE/WSH/10.01/10/Rev/1. World Health Organization, Geneva.

Yang Ch.Y. (1998) Calcium and magnesium in drinking water and risk of death from cerebrovascular disease. *Stroke* 29, 411–414.

Yang Ch.Y. & Chiu H.F. (1999b) Calcium and magnesium in drinking water and risk of death from hypertension. *American Journal of Hypertension* 12, 894–899.

Yang Ch.Y., Cheng M.F., Tsai S.S. & Hsieh Y.L. (1998) Calcium, magnesium, and nitrate in drinking water and gastric cancer mortality. *Japanese Journal of Cancer Research* 89, 124–130.

Yang Ch.Y., Chiu H.F., Cheng M.F., Hsu T.Y., Cheng M.F. & Wu. T.N. (2000) Calcium and magnesium in drinking water and the risk of death from breast cancer. *Journal of Toxicology and Environmental Health, Part A* 60, 231–241.

Yang Ch.Y., Chiu H.F., Cheng M.F., Tsai S.S., Hung Ch.F. & Lin M.Ch. (1999c) Esophageal cancer mortality and total hardness levels in Taiwan's drinking water. *Environmental Research, Section A* 81, 302–308.

Yang Ch.Y., Chiu H.F., Cheng M.F., Tsai S.S., Hung Ch.F. & Tseng Y.T. (1999a) Magnesium in drinking water and risk of death from diabetes mellitus. *Magnesium Research* 12, 131–137.

Yang Ch.Y., Chiu H.F., Cheng M.F., Tsai S.S., Hung Ch.F. & Tseng Y.T. (1999d) Pancreatic cancer mortality and total hardness levels in Taiwan's drinking water. *Journal of toxicology and environmental health, Part A* 56, 361–369.

Yang Ch.Y., Chiu H.F., Chiu J.F., Tsai S.S. & Cheng M.F. (1997) Calcium and magnesium in drinking water and risk of death from colon cancer. *Japanese Journal of Cancer Research* 88, 928–933.

Yang Ch.Y., Chiu J.F., Chiu H.F., Wang T.N., Lee Ch.H. & Ko Y.Ch. (1996) Relationship between water hardness and coronary mortality in Taiwan. *Journal of Toxicology and Environmental Health* 49, 1–9.

Yang Ch.Y., Tsai S.S., Lai T.Ch., Hung Ch.F. & Chiu H.F. (1999e) Rectal cancer mortality and total hardness levels in Taiwan's drinking water. *Environmental Research, Section A* 80, 311–316.

Yasui M., Ota K. & Yoshida M. (1997) Effects of low calcium and magnesium dietary intake on the central nervous system tissues of rats and calcium-magnesium related disorders in the amyotrophic lateral sclerosis focus in the Kii Peninsula of Japan. *Magnesium Research* 10, 39–50.

Chapter 11

Mineral water as a source of healthy minerals

Tadeusz Wojtaszek & Małgorzata Pieniak
Julian Aleksandrowicz Polish Magnesiological Society, Kraków, Poland

ABSTRACT

Minerals found in groundwater may have some healing properties for human bodies if their amount is not less than 15% of the recommended daily value. Ca and Mg are the most important. Lack of these two minerals in our daily diet can be largely offset by drinking water containing these minerals. It is estimated that a typical daily diet of the average Polish people lacks one third of these bio elements that are essential for the proper functioning of the body. Other countries face a similar problem. The healthiest mineral waters are extracted from deeper aquifers which are free of bacterial contamination and other harmful substances, and contain the appropriate level of minerals. Such waters can play a vital role in ecologically-oriented preventive healthcare programmes.

11.1   INTRODUCTION

Throughout centuries many waters have been found to have soothing, therapeutic, and preventive properties, because the minerals dissolved in waters are important for the functioning of all parts of the human body, starting from the smallest cell to entire groups of muscles, bones and internal organs (Wojtaszek, 2006).

Minerals are either leached from the rock by percolating water or they are the products of reactions occurring between various chemical components. When groundwater comes into contact with mineral deposits having components that are readily soluble in water, they are dissolved in groundwater. Particularly favorable conditions for mineral water formation occur when the CO_2 dissolved in water acidifies water causing the dissolution of less soluble minerals. Such water is highly mineralised. Gases get into the mineral waters as a result of various processes. The most important among these water-penetrating gases are: CO_2, H_2S and radon. Each type of mineral water is formed in a different way and under slightly different conditions (Ponikowska, 1996). The occurrence of various types of natural mineral waters is dependent on many factors, mainly on the geological features of the area (lithology, tectonics, paleo volcanic formations), water resources, sub-surface water flow, the age of the water and the geothermal energy (Paczyński *et al.*, 1996). Therefore, certain types of mineral waters are characteristic of different geological units.

11.2   DISCUSSION

Micro nutrients found in mineral waters are easily assimilated by the human body. Mineral water may contain up to 70 different types of components, but only ten of them are of any practical relevance when selecting the type of water suited for individual needs. These are: Mg, Ca, H_2, Cl, Na, SO_4, F, iodide, Fe, and CO_2 (Wojtaszek, 2012). The others, such as potassium, lithium, strontium, manganese, bromine, zinc and copper are present in mineral waters in such small quantities that they are practical insignificant.

11.2.1   Polish Magnesiological Association

Professor Julian Aleksandrowicz, very known Polish medical doctor and ecologist engaged in magnesium studies and a founder of the Polish Association of Magnesium called magnesium "the element of life". Due to its specific atomic structure, Mg ions have a great ability to form complexes with other substances and have unique biochemical activity. Mg is an activator for over 300 enzymes of different classes. It participates in many metabolic pathway reactions connected with the metabolism of proteins, nucleic acids, lipids and carbohydrates, and in the processes of electrolytes moving across cell membranes. It plays an important role in the division, growth and maturation of cells and in the immune reactions, reducing the course of inflammations and infections. Magnesium has a stabilising effect on cell membranes and also activates the sodium-potassium pump (Na/K-Atpase) and the calcium pump (Ca-Atpase). It is involved in the regulation of extra- and intracellular fluids as well as reactions connected with the hormone-receptor complex in cell membranes.

Another important function of Mg is activating the enzymes that are responsible for the formation, storage and use of high-energy compounds; in other words, it plays a part in regulating the energy system of the human body. Mg is a Ca antagonist – it increases the threshold level of neuromuscular excitability and shows anticonvulsant and antispastic effects in reducing muscle contraction. The adult human body contains about 24 g Mg, localised primarily intracellularly. More than 50% of the body's Mg is found in the bone tissue, about 27% in the muscles, and about 19% in other soft tissues. The total Mg concentration in the serum can be divided into three fractions: the protein bound fraction (mainly associated with albumin), the anion fraction (citrate, lactate and bicarbonate anions) and the ionized fraction (metabolically active) (Papierkowski, 2002). The recommended daily intake of Mg for an adult person is 375 mg.

11.2.2   Calcium

Ca is the main component of bones and teeth. It has a positive effect on metabolism and is essential for maintaining the normal heart function and activity of the neuromuscular system. It facilitates the healing in certain inflammatory processes, prevents osteoporosis and reduces the risk of cancer.

The daily human requirement for this bioelement is about 800–1200 mg, though it can be higher in adolescents during rapid growth, in pregnant and lactating women and in elderly people, including postmenopausal women. Calcium deficiency during the active growth phase is the cause of achieving a low peak bone mass, which can lead to the early emergence of osteoporosis. The average diet in Poland does not provide enough Ca. It is estimated that it covers only 40–80% of the required amount.

11.2.3   Bicarbonate

Bicarbonates are present in all natural mineral waters. Waters with higher levels of bicarbonates neutralise stomach acids, thus having a positive effect for people who suffer from antiacidity; at the same time they are not recommended for people with hypoacidity. Waters with high levels of bicarbonate reduce sugar content in the blood and urine, have positive effects on insulin functioning, and help regulate blood pH levels. They are physiologically active when their concentration in water is higher than 600 mg /l which is the recommended value for waters consumed as diet aids.

11.2.4   Sodium

Na is the vital component of bodily fluids and the main factor influencing the amount of water in the body. It prevents dehydration and maintains the correct acid-alkaline balance in the body. It regulates the body's electrolyte balance. Na also facilitates some of the blood functions. It is indispensable in the processes of digestion and absorption of food ingredients. Its deficiency in the body causes weakness and digestive disorders. The problem is that we tend to consume excessive amounts of Na in our diets, both in processed and home-made foods. An adult person needs up to 5 g of salt a day. Salt intake in Poland is too high with over 10 g per person per day, whereas the WHO recommended daily amount is 5 g. Water containing 250 mg-Na/l could cover only up to 5% of our body's daily requirement. Drinking waters containing Cl and Na can help supplement the body with these minerals after they have been flushed out with sweat, which also improves our mood. In such instances, it is recommended to drink waters containing from 200 to 1000 mg Na and from 250 to 1500 mg Cl/l (Wojtaszek, 2012).

11.2.5   Chlorides

Mineral waters differ considerably in their Cl content, ranging from several to over 1500 mg/l. Waters with higher Cl concentrations usually have higher Na content as well. The loss of Cl from the body distorts the natural pH levels which can lead to a variety of disorders, from dehydration to a coma. Maintaining the normal hydration level and osmotic pressure in the body is largely dependent on the alkaline content of the bodily fluids. Cl is the most important acid fraction of the plasma. Cl anions are found in the gastric juice in the form of hydrochloric acid; as sodium chloride, and it is responsible for maintaining the normal osmotic pressure of the blood and bodily fluids. Heavy physical work, sporting activities, or hot summer temperatures make our bodies sweat and flush out large amounts of Na which can cause a deficiency and consequently lead to weakness, rapid fatigue, or other dysfunction of the body. Drinking waters with high Cl and Na content on such occasions can help supplement sodium deficiencies and have a positive influence on our mood and the body's functioning. Whenever considerable amounts of Na are lost during intense physical effort, it is recommended to drink waters containing even as much as 1000 mg Na and 1500 mg Cl/l.

11.2.6   Sulphate

Sulphate mineral waters facilitate metabolism, enhance the secretory function of the liver and bile, have healing effects in inflammatory processes of the urinary tract and in chronic inflammatory conditions of the intestines. In the presence of Ca cations, sulphates influence the insulin functioning and reduce the content of sugars in the blood and urine. SO_4 waters are physiologically active when they contain at least 250 mg-SO_4/l. In the simultaneous presence of Ca cation, the activity of insulin is influenced reducing the sugar content in the blood and physiological urinary. Concentration exceeding 600 mg/l can sometimes cause diarrhea.

11.2.7   Iodine

Iodine is the essential component needed for the synthesis of thyroid hormones in the body and for the proper functioning of the thyroid gland. The daily requirement for iodine in an adult person is 150 µg. Iodine greatly contributes to the healthy pregnancy and proper development of the fetus and to the proper mental and physical development of children through youth. Insufficient amounts of iodine in the body can be the cause of thyroid disorders such as goiter or hypothyroidism. Frequency of thyroid disorders has been reported to be much higher in the areas where drinking water contains almost no iodine. Natural mineral waters containing around 0.15 mg of iodine per liter and are a good source of this mineral.

11.2.8   Fluoride

Fluoride plays a vital role in bone and teeth mineralisation. It has a noticeable effect on the body when its concentration in mineral water is 1 mg/l or more, but the excess of 1,5 mg/l has negative effect.

11.2.9   Iron

Mineral waters containing bivalent iron ions can be very healthy as iron in this form is more easily assimilated into the body than other types. Drinking such waters can have positive effects in people with anemia or it can enhance metabolism. Mineral waters containing iron salts are physiologically effective when the iron concentration is equal or more than 1 mg/l.

11.2.10   Carbon dioxide

We can distinguish between naturally and artificially carbonated waters. CO_2 in mineral water slightly irritates the mucous membrane in the mouth, has a refreshing effect, and stimulates our digestive system. For a healthy person it is better to consume medium-carbonated waters as they can enhance digestive processes and diuresis, and also help increase resorption of other ingredients in food and water. CO_2 in water has a bacteriostatic effect and acts as a preservative. Sparkling water is much healthier than still water. Waters containing more than 4000 mg-CO_2/l are not recommended for people predisposed to hyperacidity, for those who have stomach ulcers, major problems with the circulatory system, throat or vocal cords disorders, or for small children (Pieniak *et al.*, 2012).

Other minerals often listed on the mineral content label of bottled waters, such as potassium, lithium, bromine, copper, or even silver are of no practical use as their amount in mineral waters is much lower than their recommended daily values and thus they do not have any nutritional value.

11.3 CONCLUSIONS

Usually people are deficient by about 100 mg Mg and at least 200 mg Ca in their daily diet. These quantities can be easily supplemented by drinking mineral waters containing at least 50 to 100 mg Mg and 150 mg Ca/l (Pieniak *et al.*, 2012). Ca and Mg determine the hardness of water.

The importance of water hardness for human health was first described in 1950. The relationship between the hardness of water and the onset of vascular disease was first described by the Japanese chemist Kobayashi (Kobayashi, 1957) whose study showed higher mortality rates for cerebrovascular diseases in the areas with more acidic (softer) water as compared to the areas where more alkaline (harder) water was used for drinking (Kobayashi, 1957).

Chronic magnesium deficiency combined with a reduction in its concentration in the myocardium are the results of insufficient supply of this mineral in the diet. This is specifically connected with the scarcity of Mg ions in water may predispose to severe cardiac diseases (Durlach, 1984).

Epidemiological studies have established an inverse correlation between the hardness of drinking water and the death rate from cardiovascular diseases (both in adults and infants). Drinking hard water reduces cardiovascular risks, and vice versa – soft water consumption is likely to increase those risks. Mg has the strongest inverse correlation to mortality rate from cardiovascular disease (Marier, 1980; 1982; 1980).

REFERENCES

Durlach J., Bara M. & Guiet_Bara A. (1984) Taux de Mg de l, eau du boisson et facteur de risque cardiovasculaire. III. Jornadas de Alimentacaoe Dietetica de Coimbra, 29–31, Marcio 1984, Coimbra Medica, 51(5), 129–142.

Kleczkowski A.S. & Różkowski A. (Editors) (1997) *Słownik hydrogeologiczny (Hydrogeological glossary)*. Ministerstwo Ochrony Środowiska Zasobów Naturalnych i Leśnictwa, Warszawa, p. 208.

Kobayashi J. (1957) On geographical relationship between the chemical nature of river water and death rate from apoplexy. Berichte des *Ohara Institutes fur Landurtschaftuche Biologie* 11, 12–21.

Marier J.R. (1980) Role du Mg dans la cardioprotection des eaux dures. Medecine et nutrition, 16, 1, 23–29.

Marier J.R. (1982) Role of environmental MG in cardiovascular diseases. Magnesium, 1(3/6), 266–276.

Marier J.R., Neri L.C. & Anderson T.W. (1980) Durete de l,eau, sante et importance du magnesium. NRCC/CNRC n° 17582, Ottawa, p. 123.

Paczyński B. & Płochniewski Z. (1996) *Wody mineralne i lecznicze Polski, (Mineral and medicinal waters in Poland)*. PIG, Warszawa, p. 190.

Papierkowski A. (2002) *Znaczenie magnezu w praktyce lekarskiej. Część I. Przyczyny i objawy zaburzSospodarki magnezowej, (Significance of magnesium in medical practices)*. Medycyna Rodzinna 1, 31–34.

Pieniak M. & Wojtaszek T. (2012) Significance of the most important mineral components in bootled waters, Calcium and magnesium in groundwater distribution and significance. Book of abstracts and field trip guidebook. International Seminar. Katowice, Poland.

Ponikowska I. (1996) *Lecznictwo uzdrowiskowe, Poradnik dla chorych*, (Curative medicine) Oficyna Wydawnicza Brandta, Bydgoszcz. p. 83.

Wojtaszek T. (2006) *Wody mineralne służą zdrowiu (Mineral waters are healthy)*, Agro Przemysł, Nr specjalny.

Wojtaszek T. (2012) Water as the mineral nutrients carrier, Calcium and magnesium in groundwater distribution and significance. Book of abstracts and field trip guidebook. International Seminar. Katowice, Poland.

Chapter 12

Mg and Ca in groundwater and the incidence of acute coronary syndrome: Application of a Bayesian spatial method in medical geology

Anne Kousa[1], Aki S. Havulinna[2], Niina Puustinen[3] &
Elena Moltchanova[4]
[1]*Geological Survey of Finland, Kuopio, Finland*
[2]*National Institute for Health and Welfare, Helsinki, Finland*
[3]*Aurora Hospital, City of Helsinki, Department of Social Services and Health Care, Helsinki, Finland*
[4]*Department of Mathematics and Statistics, University of Canterbury, Christchurch, New Zealand*

ABSTRACT

Geographical health studies have been carried out previously using administrative areas as study units. Instead of administrative units, a 10 km by 10 km grid has now been used. A spatial hierarchical Conditional Autoregressive (CAR) model was fitted within a Bayesian framework. This chapter describes the Bayesian model formulation for the small area analyses. The association between Ca and Mg in groundwater and the incidence of acute coronary syndrome, a subclass of coronary heart disease, was estimated. In total 93 205 acronym ACS cases (67 755 men and 25 450 women) aged 35–74 years living in the rural Finland (excluding Lapland, Åland and the Turku Archipelago), were identified from the Finnish Cardiovascular Disease Register during 1991–2003. The data on the corresponding population-at-risk were obtained from the Finnish Population Register Centre. Ca and Mg concentrations were obtained from the groundwater database of the Geological Survey of Finland. The geographic pattern of acute coronary syndrome incidence was quite similar in men and women diagnosed in 1991–2003, suggesting common risk factors for both sexes and the data on men and women were, therefore, pooled. A 1 mg/l increment in Mg concentration was associated on average with a 2% (95% highest density regions: 0.3% to 3.9%) decrease in ACS incidence. Ca concentration in groundwater did not have any marked association with the incidence of ACS. Low natural concentrations of Mg in local groundwater were associated with an increased incidence of acute coronary syndrome in rural Finland.

12.1 INTRODUCTION

Medical geology is defined as the science examining the relationship between natural geological factors and health and understanding the influence of ordinary environmental factors on the geographical distribution of health problems (Selinus *et al.*, 2005). The ecological study design is commonly used in studies of medical geology and geography. In epidemiological terms, the ecological approach means that the statistical analyses are based on group level data instead of individual data. In ecological population studies, the aggregated groups are inhabitants and related data from some mutually exclusive geographic areas represented by area level data (Gatrell, 2002). The size of the study unit is a key issue in ecological studies, due to the Modifiable Areal Unit Problem (MAUP) (Openshaw, 1984). Geographical health studies have been carried out previously using administrative regions, such as municipalities, provinces, communes, cities, hospital districts or postal areas as study units. Analysing the data such units may not be suitable for mapping health outcomes and the choice of boundaries may have a major influence on the results (Jarup, 2004). Finland is an example of a sparsely inhabited country with a geographically uneven population distribution. The majority, approximately 75%, of the population, lives in an area that covers 5% of the entire land area of the country. Half of the population lives in the cities (Rusanen *et al.*, 1995). Administratively defined areas are very heterogeneous in size and number of population-at-risk in Finland. Therefore, the interpretations of the regional variation in the occurrence of diseases are challenging. In this study, using the regular grid-based boundaries helped to reduce the MAUP. If the regions are too small with only a few cases, distorted spatial patterns may be observed due to random variation (Ranta *et al.*, 1999). The Bayesian smoothing technique used in this study is a valuable tool for reducing the effect of random variation in the results (Ranta *et al.*, 1999). The aim of this chapter is to describe the Bayesian statistical method used in an earlier published medical geology study (Kousa *et al.*, 2008).

The incidence of Acute Coronary Syndrome (ACS) has been among the highest in the world in Finland (Pajunen *et al.*, 2004). Several epidemiological studies have shown geographical variation in the occurrence of Coronary Heart Disease (CHD) worldwide and also within Finland (Figure 12.1). (Näyhä, 1989; Jousilahti *et al.*, 1998; Karvonen *et al.*, 2002; Havulinna *et al.*, 2008). Mortality and morbidity due to CHD are higher in eastern Finland compared with western part of the country. Despite declining countrywide trends (Pajunen *et al.*, 2004) the regional differences in Cardio-Vascular Disease (CVD) mortality and morbidity have remained almost unchanged for over 60 years (Kannisto, 1947; Karvonen *et al.*, 2002; Havulinna *et al.*, 2008). Low socioeconomic status is associated with higher CHD mortality and morbidity rates (Hallqvist *et al.*, 1998). However, the socioeconomic status does not explain the east-west difference in the incidence of ACS and coronary mortality rates (Salomaa *et al.*, 2000). Association of drinking water constituents with the occurrence of CVD have been evaluated in several studies in different parts of the world. Many of these studies give conflicting results and therefore no consensus on this subject has been reached so far (see reviews by Monarca *et al.*, 2006; Catling *et al.*, 2008).

Finnish groundwater is slightly acidic and very soft, with low concentrations of Ca and Mg. There is, however, a regional variation of the concentration of these elements. Regionally high Ca and Mg concentrations are found in the western part of the country and in the southern coastal areas (Figure 12.1). Clayey and silty sediments and carbonate rocks increase water hardness in these areas. The regional distribution of water hardness closely resembles those of Ca and Mg. The soft water areas are found in sand and gravel deposits in glaciofluvial formations (Tarvainen *et al.*, 2001). In rural Finland, the most common well type is a dug well with concrete rings. The most common soil type is glacial till. Private dug wells are generally 3–10 m deep and 80–120 cm in diameter. Wells drilled in bedrock are generally 40–80 m deep and

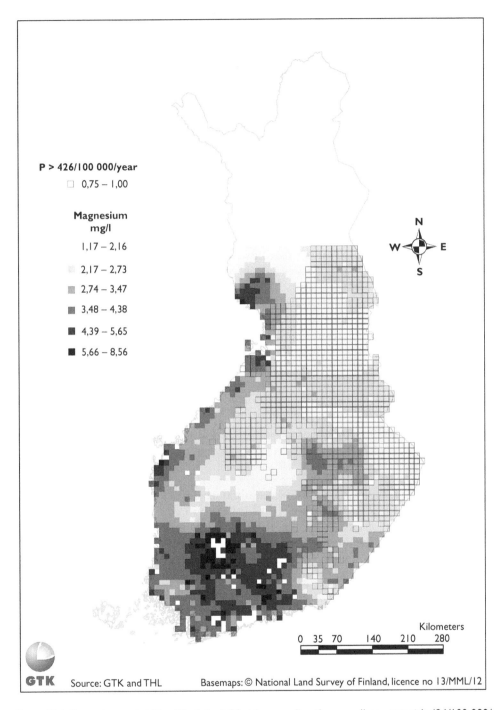

P > 426/100 000/year

☐ 0,75 – 1,00

Magnesium
mg/l

1,17 – 2,16

2,17 – 2,73

2,74 – 3,47

3,48 – 4,38

4,39 – 5,65

5,66 – 8,56

Kilometers

0 35 70   140   210   280

GTK   Source: GTK and THL   Basemaps: © National Land Survey of Finland, licence no 13/MML/12

Figure 12.1 Posterior probability (P) of the ACS risk exceeding the overall country risk 426/100 000/
year among 35–74 year old men and women (pooled) in 1991–2003 (square symbol).
Regional distribution of Mg (mg/l) concentration in local groundwater in the rural Finland
(grey-scale). White squares: outside the target area of the study.

119 mm in diameter (Lahermo *et al.*, 1990). The median concentrations in dug wells are 11.4 mg-Ca/l and 2.4 mg-Mg/l. In wells drilled in bedrock the median values are slightly higher: 16 mg-Ca/l and 4.5 mg-Mg/l (Tarvainen *et al.*, 2001). These levels are below the average in Europe. In European surface waters the median values of Ca and Mg are 40 mg/l and 6.02 mg/l, respectively (Salminen *et al.*, 2005). The relation of the main components of water hardness, Ca and Mg, with the incidence of ACS was estimated with separate ecological regression models.

12.2 METHODS

12.2.1 Study subjects and population at risk

A total of 93 205 ACS cases, 67 755 men and 25 450 women, aged 35–74 years living in the rural Finland (excluding Lapland, Åland and the Turku Archipelago), were obtained from the Finnish Cardiovascular Disease Register (CVDR). The data of occurrence of CHD and cerebrovascular diseases from 1991–2003 were available at the time of the study (Laatikainen *et al.*, 2004). Data on the first ACS events leading to hospitalisation or death were included only in the analysis, and thus soft CHD is excluded. Each permanent resident in Finland is assigned a unique personal Identification Number (ID) which enables accurate computerised record linkage. Using the personal ID code, each person with ACS was located by the map coordinates of the place of residence at the time of diagnosis. The corresponding aggregated data on population at risk were obtained from Statistics Finland. Based on the regional information (Keränen *et al.*, 2000) the analyses was restricted to rural areas only. The details of ACS case definitions were presented by Kousa *et al.* (2008).

12.2.2 Geochemical data

The concentrations of Ca and Mg were obtained from the groundwater database of the Geological Survey of Finland (GTK) (Lahermo *et al.*, 1990; Tarvainen *et al.*, 2001). The concentrations of certain elements in the groundwater were determined using Inductively Coupled Plasma Mass Spectrometry (ICP–MS) and Inductively Coupled Plasma Atomic Emission Spectrometry (ICP–AES) methods. Over 4300 water samples were collected, mainly from dug wells and wells drilled in bedrock, and analysed between 1992 and 2005. In statistical analyses, the geochemical data for dug wells and drilled wells were pooled. Ca and Mg measurements were interpolated into 10 km by 10 km grid cells (Figure 12.1). The description of the interpolation model was presented by Kousa *et al.* (2008, Appendix A).

12.2.3 Bayesian statistical methods

A spatial hierarchical conditional autoregressive model was fitted within a Bayesian framework. Instead of administrative boundaries, the geo-coded data were aggregated into 10 km by 10 km grid cells. The relatively large grid cell size ensures the protection of privacy of individuals. The geographic pattern of ACS incidence was quite similar in men and women who were diagnosed in 1991–2003, suggesting common risk factors for both sexes. Therefore, data on men and women were pooled.

A Bayesian conditional autoregressive model which has been developed by Besag *et al.* (1991) is now widely used in disease mapping. In a wider scope, it can be described as a generalised linear model with Poisson link and spatially auto-correlated residuals. The formulation as best fitted to this specific problem is as follows:

Let Y_{ik} denote the number of cases observed in the area i, in the age-group k. It is common to model such counts using Poisson distribution with the mean proportional to the area- and age-group specific risk μ_{ik}, as well as to the population-at-risk of that age-group in the area, N_{ik}. The area- and age-specific risk, in turn depends on the global risk e^{α}, the age-group specific relative risk $e^{h(k)}$, local area-specific environmental covariates Z_i and area-specific residual λ_i. This dependency can be described with the following log-linear regression equations:

$$Y_{ik} \sim Poisson(\mu_{ik}N_{ik})$$

$$\log(\mu_{ik}) = \alpha + h(k) + Z_i\xi + \lambda_i$$

where ξ is the vector of regression coefficients, representing the effect of environmental covariates Z_i on the risk. The most general form of the age-group specific risk is non-parametric

$$h(k) = \beta_k$$

Depending on the nature of the disease, any functional form may be used instead. For-example, a risk proportional to age would be

$$h(k) = \beta k$$

The spatial component is taken into account via the residual λ_i. While in the standard regression model, these terms are assumed to be a sample of independent identically distributed normal variables with mean 0, here they are assigned an auto-correlated spatial prior:

$$\lambda_i \sim N(\overline{\lambda_{-i}}, \tau m_i)$$

Here, $\overline{\lambda_{-i}}$ is the mean of λ-terms in the neighbourhood of i, m_i is the number of neighbours in that neighborhood, and τ is the spatial precision (= inverse variance) parameter. Note that the standard Bayesian notation of $N(mean, \tau = \frac{1}{sd^2})$ is used rather than the more common $N(mean, sd^2)$. Smaller values of τ refer to smaller precision, i.e. greater spatial variation which is unexplained by either age or environmental covariates included in the model.

The definition of the neighborhood is commonly taken to include all the areas adjacent through either side or corner (Figure 12.2). However, other configurations are possible. For example, when modelling contagious diseases, the connectivity of

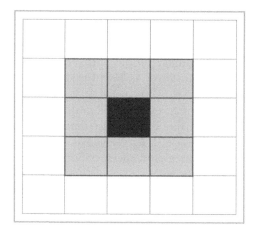

Figure 12.2 Neighbors are defined as the grid cells that are adjacent by a side or a corner.

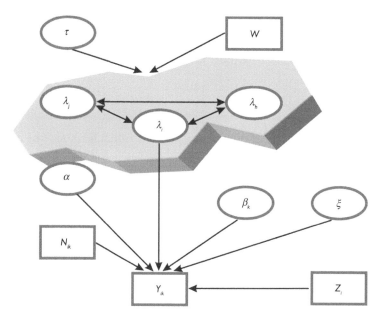

Figure 12.3 The Directed Acyclic Graph (DAG) of the regression model used in this study. W is the neighbourhood indicator: $W_{ij} = 1$, if the grid cells i and j are neighbours, and $W_{ij} = 0$ otherwise. λ_j and λ_h represent the area-specific log-relative risk in the grid cells which are neighbours to the grid cell i. Other variables are explained in the text.

regions (as measured by time required to travel from one area to another) rather than actual geographic distances may be a better guide for defining neighborhoods.

Standard non-informative priors are then assigned to the parameters. For the regression coefficients α, β, and ξ that means $N(0, .00001)$ and for the spatial precision τ, a Gamma(.01, .01) prior. A directed acyclic graph (DAG) (Figure 12.3.) provides an overall view of the model.

The model was estimated using WinBUGS (Spiegelhalter *et al.*, 2004) and the results were plotted using ArcGIS 9.3.1.

12.3 RESULTS

The age-adjusted ACS incidence was 589/100 000/year (95% HDR 584–593) in men and 177/100 000/year (95% HDR 175–179) in women aged 35–74 years in rural Finland (Kousa *et al.*, 2008). The basic descriptive statistics of interpolated geochemical data are provided in Table 12.1. Age group, sex, and Mg or Ca were included as covariates in the spatial model. A 1 mg/L increment in Mg concentration was associated on average with a 2% lower incidence of ACS. Ca concentration did not have any notable association with the incidence of ACS (Table 12.2).

12.4 DISCUSSION

An inverse association between the incidence of ACS and Mg in local groundwater was found in this study (Kousa *et al.*, 2008). The results support previous findings about the protective role of Mg in groundwater against CHD incidence (Rubenowitz *et al.*, 2000; Kousa *et al.*, 2006; Monarca *et al.*, 2006).

A recent study suggests an association between supplemental calcium intake and an increased risk of cardiovascular disease in men (Xiao *et al.*, 2013). However, our findings suggest that the differences in calcium intake from drinking water might not be so large that they would affect the risk of ACS.

The geographic coordinate based data can be presented as thematic maps where the individual data have been aggregated into regular grids, such as 10 km × 10 km without the need to use administrative units. An additional advantage in Finland is that there are uniform health data from nationwide Causes of Death and Hospital Discharge Registers which in practice cover 100% of events and the diagnostic

Table 12.1 Interpolated geochemical data for local groundwater (after Kousa *et al.*, 2008).

Element mg/l	Median	Mean	SD	Min	Max
Mg	2.70	3.03	1.13	1.17	10.23
Ca	12.35	13.33	4.06	5.17	40.85

Table 12.2 The estimated effects of Mg and Ca on the incidence of ACS among men and women (pooled) in 1991–2003 in rural Finland (after Kousa *et al.*, 2008).

Element mg/l	Posterior mean(%)*	95% HDR
Mg**	−2.2	−3.9; −0.3
Ca	−0.1	−0.6; 0.4

The regression coefficient were estimated in a single spatial model.
*Adjusted for age group and sex.
**95% HDR does not include zero.

classification have been shown to be of high quality (Pajunen *et al.*, 2005; Tolonen *et al.*, 2007). The unique personal ID enables accurate linkage of records, such as geographic coordinates and disease data, between different data sources.

It is important to note some limitations of the present ecological study. First, the individual exposure levels were not measured. The exposure data, Ca and Mg in groundwater, were aggregated into $10\,km \times 10\,km$ grids based on the chemical analyses of the water sampled from over 4000 of dug wells and bedrock wells, and these aggregated estimates were used as covariates in the present analysis. Also, by being constituents of water hardness, the Ca and Mg, concentrations are correlated. Second, the findings were controlled for age and gender but not for potential confounding factors such as classical CHD risk factors like hypertension or serum cholesterol. These risk factors, unlike the concentration of environmental elements, should be known preferably at the individual level and they are not available for disease cases in the register, let alone in the whole at-risk population. Third, it was not possible to determine whether some part of the population which was defined to be rural was in fact served by a public water supply. The results from ecological studies do not justify causality but are useful for generating and exploring hypotheses (Rothman, 1993).

In conclusion, soft groundwaters with low Mg concentrations were associated with an increased incidence of ACS in the rural Finland. The geographical pattern of the high incidence of ACS may reflect common risk factors which are concentrated in eastern Finland. Further studies are needed to prove- or disprove whether the low Mg concentration in drinking water is associated with the risk of CHD.

REFERENCES

Besag J., York J. & Mollié A. (1991) Bayesian Image Restoration, with two applications in spatial statistics. *Annals of the Institute of Statistical Mathematics* 43(1), 1–21.

Catling L.A., Abubakar I., Lake I.R., Swift L. & Hunter P.R. (2008) A systematic review of analytical observational studies investigating the association between cardiovascular disease and drinking water hardness. *Journal of Water and Health* 6(4), 433–442.

Gatrell A.C. (2002) *Geographies of health: An introduction*. Blackwell Publishing, UK.

Hallqvist J., Lundberg M., Diderichsen F. & Ahlbom A. (1998) Socioeconomic differences in risk of myocardial infarction 1971–1994 in Sweden: time trends, relative risks and population attributable risks. *International Journal of Epidemiology* 27, 410–415.

Havulinna A.S., Pääkkönen R., Karvonen M. & Salomaa V. (2008) Geographic Patterns of Incidence of Ischemic Stroke and Acute Myocardial Infarction in Finland During 1991–2003. *Annals of Epidemiology* 18(3), 206–213.

Jarup L. (2004) Health and environment information systems for exposure and disease mapping, and risk assessment. *Environmental Health Perspectives* 112(9), 995–997.

Jousilahti P., Vartiainen E., Tuomilehto J., Pekkanen J. & Puska P. (1998) Role of known risk factors in explaining the difference in the risk of coronary heart disease between eastern and south-western Finland. *Annals of Medicine* 50, 481–487.

Kannisto V. (1947) *The causes of death as demographical factors in Finland*. [In Finnish, English summary]. Helsinki. Kansantaloudellisia tutkimuksia-Economic studies XV.

Karvonen M., Moltchanova E., Viik-Kajander M., Moltchanov V., Rytkönen M., Kousa A. & Tuomilehto J. (2002) Regional Inequality in the Risk of Acute Myocardial Infarction in Finland: A Case Study of 35- to 74-Year-Old Men. *Heart Drug* 2, 51–60.

Keränen H., Malinen P. & Aulaskari O. (2000) Suome n maaseutututyypit. [In Finnish]. Research Papers 20. Finnish Regional Research. Sonkajärvi, Finland.

Kousa A., Havulinna A.S., Moltchanova E., Taskinen O., Nikkarinen M., Eriksson J. & Karvonen M. (2006) Calcium to magnesium ratio in local ground water and incidence of acute myocardial infarction among males in rural Finland. *Environmental Health Perspectives* 114, 730–734.

Kousa A., Havulinna A.S., Moltchanova E., Taskinen O., Nikkarinen M., Salomaa V. & Karvonen M. (2008) Magnesium in well water and the spatial variation of AMI incidence in rural Finland. *Applied Geochemistry* 23, 632–640.

Laatikainen T., Pajunen P., Pääkkönen R., Keskimäki I., Hämäläinen H., Rintanen H., Niemi M., Moltchanov V. & Salomaa V. (2004) National Cardiovascular Disease Register, statistical database. Available from www.thl.fi/cvdr.

Lahermo P., Ilmasti M., Juntunen R. & Taka M. (1990) *The Geochemical Atlas of Finland, Part 1.* The hydrogeochemical mapping of Finnish groundwater. Geological Survey of Finland, Espoo.

Monarca S., Donato F., Zerbini I., Calderon R.L. & Creau G.F. (2006) Review of epidemiological studied on drinking water hardness and cardiovascular diseases. *European Journal of Cardiovascular Prevention and Rehabilitation* 13, 495–506.

Näyhä S. (1989) Geographical variation in cardiovascular mortality in Finland 1961–1985. *Scandinavian Journal of Social Medicine – Supplementum* 40, 1–48.

Openshaw S. (1984) Norwich: Geo Books. ISBN 0-86094-134-5.

Pajunen P., Koukkunen H., Ketonen M., Jerkkola T., Immonen-Räihä P., Kärjä-Koskenkari P., Mähönen M., Niemelä M., Kuulasmaa K., Palomäki P., Mustonen J., Lehtonen A., Arstila M., Vuorenmaa T., Lehto S., Miettinen H., Torppa J., Tuomilehto J., Kesäniemi Y.A., Pyörälä K. & Salomaa V. (2005) The validity of the Finnish Hospital Discharge Register and Causes of Death Register data on coronary heart disease. *European Journal of Cardiovascular Prevention and Rehabilitation* 12(2), 132–137.

Pajunen P., Pääkkönen R., Juolevi A., Hämäläinen H., Keskimäki I., Laatikainen T., Moltchanov V., Niemi M., Rintanen H. & Salomaa V. (2004) Trends in fatal and non-fatal coronary heart disease events in Finland during 1991–2001. *Scandinavian Cardiovascular Journal* 38, 340–344.

Ranta J. (2001) On probabilistic models for surveillance and prediction of disease incidence with latent process: case studies on meningococcal outbreaks, childhood diabetes and poliomyelitis. Academic dissertation. Faculty of Science. University of Helsinki. Helsinki. pp. 32.

Ranta J., Rytkönen M. & Karvonen M. (1999) A few aspects on spatially referenced data, health geographics and statistical methods. In Finnish. English summary. *Journal of Social Medicine* 36, 285–293.

Rothman K.J. (1993) Methodologic frontiers in environmental epidemiology. *Environmental Health Perspectives* 101(Suppl 4), 19–21.

Rubenowitz E., Molin I., Axelsson G. & Rylander R. (2000) Magnesium in drinking water in relation to morbidity and mortality from acute myocardial infarction. *Epidemiology* 11, 416–421.

Rusanen J., Räisänen S., Naukkarinen A. & Colpaert A. (1995) Definition of rural areas based on primary production by means of GIS. *Terra* 107, 101–111.

Salminen R. (editor.), Batista M.J., Bidovec M., Demetriades A., De Vivo B., De Vos W., Duris M., Gilucis A., Gregorauskiene V., Halamic J., Heitzmann P., Lima A., Jordan G., Klaver G., Klein P., Lis J., Locutura J., Marsina K., Mazreku A., O'Connor P.J., Olsson S.Å., Ottesen R.T., Petersell V., Plant J.A., Reeder S., Salpeteur I., Sandström H., Siewers U., Steenfelt A. & Tarvainen T. (2005) *Geochemical Atlas of Europe.* Espoo: Geological Survey of Finland. pp. 525.

Salomaa V., Niemelä M., Miettinen H., Ketonen M., Immonen-Räihä P., Koskinen S., Mähönen M., Lehto S., Vuorenmaa T., Palomäki P., Mustaniemi H., Kaarsalo E., Arstila M.,

Torppa J., Kuulasmaa K., Puska P., Pyörälä K. & Tuomilehto J. (2000) Relationship of socioeconomic status to the incidence and prehospital, 28-day, and 1-year mortality rates of acute coronary events in the FINMONICA Myocardial Infarction Register Study. *Circulation* 101, 1913–1918.

Selinus O., Alloway B., Centeno J.A., Finkelman R.B., Fuge R., Lindh U. & Smedley P. (eds.). (2005) *Essentials of Medical Geology. Impacts of the Natural Environment on Public Health*. Elsevier. Academic Press. pp. 812.

Spiegelhalter D., Thomas A., Best N. & Lunn D. (2004) WinBUGS Version 1.4.1 User Manual. MRC Biostatistics Unit, Cambridge, UK (accessed 29.4.2013). Available from: http://www.mrc-bsu.cam.ac.uk/bugs/.

Tarvainen T., Lahermo P., Hatakka T., Huikuri P., Ilmasti M., Juntunen R., Karhu J., Kortelainen N., Nikkarinen M. & Väisänen U. (2001) Chemical composition of well water in Finland – main results of the 'One thousand wells' project. In Autio. S (ed.) *Geological Survey of Finland, Current Research 1999–2000*, Special Paper 31, 57–76.

Tolonen H., Salomaa V., Torppa J., Sivenius J., Immonen-Räihä P. & Lehtonen A. (2007) The validation of the Finnish Hospital Discharge Register and Causes of Death Register data on stroke diagnoses. *European Journal of Cardiovascular Prevention and Rehabilitation* 14(3), 380–385.

Xiao Q., Murphy R.A., Houston D.K., Harris T.B., Chow W.H. & Park Y. (2013) Dietary and Supplemental Calcium Intake and Cardiovascular Disease Mortality. The National Institutes of Health–AARP Diet and Health Study. *JAMA Internal Medicine* 2013, 1–8.

Part 3

Calcium and magnesium in mineral and therapeutic waters

Chapter 13

Ca and Mg in bottled mineral and spring waters in Europe

Lidia Razowska-Jaworek
Polish Geological Institute-National Research Institute,
Upper Silesian Branch, Sosnowiec, Poland

ABSTRACT

The chemistry of mineral waters is determined by the composition of the rock environment it is abstracted from. The main source of Ca in groundwaters is dissolution of minerals like: calcite, dolomite, aragonite, gypsum, fluorite, feldspars as well as reverse ion exchange. The main source of Mg is dissolution of dolomite and calcite, magnesite, ferromagnesian silicates, reverse ion exchange and dedolomitisation. In 2010 the number of registered mineral water brands in Europe was 1916. This chapter presents the results of study on chemical composition of 1903 different types of mineral waters originating from 41 European countries. Calcium concentration in these waters ranges from 0.05 to 1134 mg/l; 42% of waters show low Ca content (50–150 mg/l) and 40% very low Ca content (<50 mg/l). Magnesium concentration ranges from 0.02 to 1060 mg/l; 33% of waters show low (<30 mg/l) and 37% very low (<10 mg/l) Mg content. According to EU Directive 80/777/EEC, only 18% may be called 'mineral waters containing calcium' with Ca > 150 mg/l and only 21% may be called as 'mineral waters containing magnesium' with Mg > 50 mg/l. This is an alarming situation as the most important constituents of mineral waters for 75% of bottled mineral and spring waters in Europe are observed in extremely low concentrations. This fact, to a great extent, influences human health and well-being and should be further studied in detail.

13.1 INTRODUCTION

'Natural Mineral Water' is microbiologically wholesome water from an underground aquifer tapped at spring, natural or drilled wells (European Directive 80/777/EEC). The only treatment allowed prior to bottling is to remove unstable components such as iron and sulphides and to (re)introduce carbon dioxide. Water from a natural source that contains few minerals is called 'Spring Water'. In contrast to mineral water which has to be bottled at the source, Spring Water may be transported before bottling. For the needs of this chapter, Natural Mineral Water and Spring Water have been brought together in one data set which has been called 'mineral waters'.

The chemical composition of mineral waters is determined by the composition of the rock environment it is abstracted from and depends on the depth of an aquifer, residence time and the exposure to meteoric waters. The main source of Ca in groundwaters is dissolution of minerals: calcite, dolomite, aragonite, gypsum, fluorite, feldspars as well as reverse ion exchange. The main source of Mg is dissolution of dolomite and calcite, magnesite, ferromagnesian silicates, reverse ion exchange and dedolomitisation.

Mg is the most important water component, hence, buying bottled Natural Mineral Water we should pay attention to the Mg concentration as we need about 300 mg/day of Mg and magnesium-rich water is an important dietary source. Ca is the second important water component. As the Recommended Dietry Allowance (RDA) of Ca is 800 to 1000 mg/day, water is not the essential dietary source of calcium, but it may be very important for people who cannot consume dairy products such as milk or cheese.

13.2   APPROACH AND METHODS

In 2010 the number of registered mineral water brands in Europe was 1916. This chapter presents the results of study on chemical composition of 1903 different types of mineral waters, originating from 41 European countries (Figure 13.1). The analysed waters are commercially available from the sellers and the analyses mainly come from the web pages of the mineral water sellers as well as from the internet pages: *http://www.mineralwaters.org; http://www.efbw.eu; http://www.finewaters.co; www.zdrowawoda.pl.*

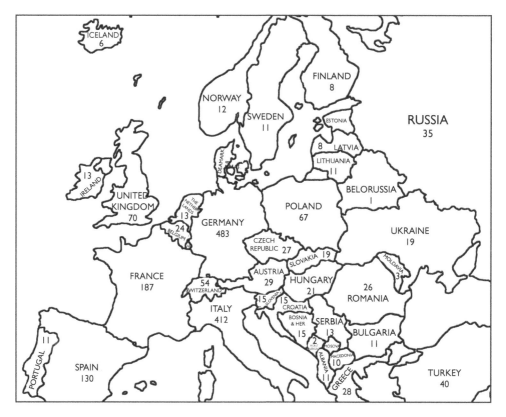

Figure 13.1 Number of examined bottled mineral and spring water brands in each European country.

The analysed mineral and spring waters come from the following countries: Albania (11), Andorra (2), Austria (29), Belarus (1), Belgium (24), Bosnia and Herzegovina (15), Bulgaria (11), Croatia (15), Cyprus (3), Czech Republic (27), Denmark (8), Finland (8), France (187), Germany (483), Greece (28), Hungary (21), Iceland (6), Ireland (13), Italy (412), Latvia (8), Lithuania (11), Luxembourg (8), Macedonia (10), Malta (2), Moldova (3), Montenegro (2), the Netherlands (13), Norway (12), Poland (67), Portugal (31), Romania (26), Russia (35), Serbia (13), Slovakia (19), Slovenia (15), Spain (130), Sweden (11), Switzerland (54), Turkey (40), Ukraine (19) and United Kingdom (70).

Consumption of bottled waters in the European Union varies from one country to another with the average consumption at 105 l/year. The lowest is in Finland, 16 l/year per inhabitant, and the highest in Italy, nearly 200 l/year per inhabitant.

13.3   RESULTS AND DISCUSSION

13.3.1   Mineral content in European mineral waters

The amount of minerals dissolved in water is indicated as Total Dissolved Solids (TDS) which is normally made of carbonates, bicarbonates, chlorides, sulphates, phosphates, nitrates, calcium, magnesium, sodium, potassium, iron, manganese, and a few other minerals. TDS is the second most important factor in matching water with food. The higher the mineral content, the more distinct a water's taste. Regulations regarding TDS in drinking waters vary throughout the world. In the United States, bottled water must contain at least 250 mg/l TDS to be labelled as mineral water. TDS above 500 mg/l qualifies as a water with 'low mineral content'; more than 1500 mg/l allows a 'high mineral content' label. High TDS could impact taste in drinking water while water of very low TDS may be unacceptable because of a flat and uninspiring taste (Bruvold & Ongerth, 1969). In the European Union there is no regulation regarding maximum levels of TDS, the only regulation is the level of electric conductivity of water which is 2500 μS/cm, equivalent to around 2000 mg/l TDS.

The natural background level of the TDS in groundwaters is 200 to 500 mg/l (Witczak et al., 2013). The studied European bottled waters fall within a TDS range of 4 to 86 550 mg/l. The average value is 1048 mg/l and the median value is 332 mg/l. Most of the examined waters (62.6%) are of low and very low mineral content with the TDS values below 500 mg/l (Figure 13.2), 20.6% are of intermediate mineral content with a TDS range 500–1500 mg/l and 8.7% are of very low mineral content (TDS < 50 mg/l). Only 16.7% are of high mineral content with a TDS above 1500 mg/l.

Waters with very low mineral content have minimal concentrations of Ca and Mg (Table 13.1). Median of Ca which is 3.5 mg/l and median of Mg which is 0.9 mg/l are definitely low and consumers should avoid such mineral waters. Waters with low mineral content, which prevail in Europe (53.9%) also have too low Ca and Mg content. The maximum values are acceptable, but medians of Ca (47.0 mg/l) and Mg (9.0 mg/l) are below recommended levels for mineral waters. Waters with mineral content ranging between 500 and 1000 mg/l have optimal levels of Ca and Mg for drinking purposes, but in bottled mineral waters consumers would expect higher values. The best contents of Ca and Mg are in waters with the TDS values within a range

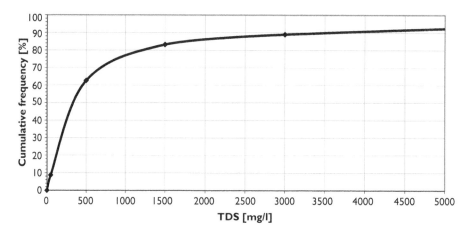

Figure 13.2 Cumulative frequency of the TDS in European bottled mineral waters.

Table 13.1 Ca and Mg in European mineral waters of different TDS.

Mineral content	TDS mg/l	Ca range median mg/l	Mg range median mg/l	Ca/Mg median	Mineral waters in Europe %
High	>1500	8.0–1134.0 222.0	1.2–1060.0 70.0	1.8	16.7
Intermediate	1000–1500	2.5–413.9 163.0	0.3–150.9 45.0	2.1	5.4
	500–1000	2.2–331.0 103.0	0.1–144.0 27.0	2.2	14.6
Low	50–500	0.05–162.4 47.0	0.04–89.5 9.0	2.4	53.9
Very low	0–50	0.3–14.0 3.5	0.02–6.8 0.9	2.5	8.7

1000 to 1500 mg/l. Ca median 163 mg/l and Mg median 45 mg/l are the most favorable. Unfortunately, only 5.4% of European mineral waters fall within this range of the TDS. These are mostly mineral water brands from Italy (30%) and some from Hungary (10%), Switzerland (10%), and France (10%) (Figure 13.3).

Calcium and magnesium contents in the most European countries are presented in Table 13.2. The table contains the median and average values of Ca and Mg concentrations and the percentage of waters falling within ranges of Ca and Mg contents selected basing on the values which are important for human health.

13.3.2   Ca in mineral waters in Europe

The main source of Ca in groundwaters is a dissolution of carbonate and sulphate minerals such as: calcite, dolomite, aragonite, gypsum, fluorite as well as anorthite,

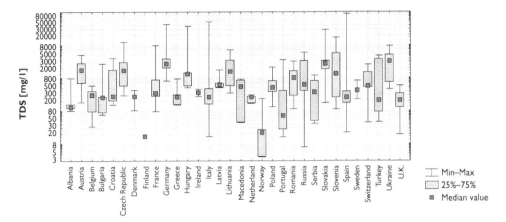

Figure 13.3 Box and Whisker plot of the TDS of bottled mineral waters in different European countries.

diopside and amphibolite during weathering processes. In deep aquifers the reverse ion exchange may be the additional source of calcium in waters. Calcium content in groundwaters is controlled by carbonate equilibrium and strongly depends on CO_2 and pH (Hem, 1989). The most abundant Ca speciations in groundwaters are: Ca^{2+}, $CaSO_4^0$, $CaHCO^{3+}$. Natural background level of Ca in groundwater is 2–200 mg/l (Witczak *et al.,* 2013).

Calcium concentration in the European waters that were examined ranges from 0.05 to 1134 mg/l, average is 111 mg/l and median is 70 mg/l. The values of Ca are generally higher than the values of Mg. Ca values show significant difference among the brands. 42% of European mineral waters show low Ca content (50–150 mg/l) and 40% very low Ca content below 50 mg/l (Figure 13.4). According to EU Directive 80/777/EEC only 25% of waters may be called 'mineral waters containing calcium' with Ca content above 150 mg/l.

Waters with the preferable content of Ca occur in countries of central Europe such as: Austria, Czech Republic, Hungary, Lithuania, Slovakia, Slovenia and Switzerland (Figure 13.5). The lowest values of Ca are observed in waters from Albania, Bulgaria, Greece, Portugal, Norway, Serbia and the UK.

The optimum concentration of Ca in mineral waters is above 150 mg/l. Only in 10 countries do more than 30% of bottled mineral waters have such high calcium content (Figure 13.6). These are water brands from: Slovakia, Hungary, Austria, Lithuania, Macedonia, Romania, Czech Republic, Bosnia and Herzegovina, Germany and Switzerland.

In some countries none of the bottled mineral waters have Ca higher than 150 mg/l. These are the 10 countries: Belgium, Croatia, Denmark, Finland, Iceland, the Netherlands, Norway, Portugal, Sweden and the UK.

Despite its beneficial effects on the human body, Ca overdose causes some negative effects. An excessive amount of calcium in the body is called a hypercalcemia. This condition can impair kidney function; sometimes even cause kidney stones. Hypercalcemia can cause constipation and reduces the absorption of other important

Table 13.2 Calcium and magnesium in mineral and spring waters in European countries.

Country	No of waters	Ca mg/l		Ca range (mg/l) %				Mg mg/l		Mg range (mg/l) %			
		Median	Average	<10	10–50	50–150	>150	Median	Average	<10	10–30	30–100	>50
Albania	11	48.1	73.8	9.1	63.6	18.2	9.1	5.2	7.9	54.5	36.4	0.0	0.0
Austria	29	146.0	157.2	0.0	17.2	34.5	48.3	37.2	41.8	10.3	24.1	55.2	17.2
Belgium	24	62.5	54.9	25.0	20.8	54.5	0.0	7.2	10.1	54.2	37.5	4.2	0.0
Bosnia&Her	15	90.2	147.7	0.0	0.0	60.0	33.3	44.3	89.7	33.3	6.7	26.7	33.3
Bulgaria	11	7.2	32.1	63.6	18.2	9.1	9.1	5.6	14.0	27.3	9.1	9.1	0.0
Croatia	15	58.4	58.8	0.0	33.3	60.0	0.0	16.6	18.9	20.0	60.0	20.0	0.0
Czech Rep.	27	75.8	119.1	7.4	29.6	22.2	37.0	30.5	85.5	11.1	37.0	25.9	33.3
Denmark	8	50.0	51.2	12.5	25.0	50.0	0.0	6.0	7.6	100.0	0.0	0.0	0.0
Finland	8	3.5	6.1	75.0	25.0	0.0	0.0	1.5	2.3	87.5	0.0	0.0	0.0
France	187	71.0	97.8	17.6	18.2	43.8	20.3	13.1	26.2	38.5	33.7	25.1	15.5
Germany	483	100.0	159.2	4.5	17.0	44.9	33.5	27.7	52.3	23.0	28.4	39.5	30.0
Greece	28	50.3	63.0	7.1	42.8	42.8	7.1	12.3	18.5	32.1	46.4	21.4	7.1
Hungary	21	156.5	163.0	4.7	0.0	28.6	52.4	46.4	200.2	4.8	28.6	57.1	47.6
Iceland	6	5.5	4.9	100.0	0.0	0.0	0.0	1.7	1.8	100.0	0.0	0.0	0.0
Ireland	13	101.0	91.3	15.4	7.7	61.5	15.4	16.0	16.3	30.8	61.5	7.7	0.0
Italy	412	57.3	94.2	10.7	32.3	42.2	12.9	13.0	24.0	42.2	33.5	19.7	11.4
Latvia	8	61.2	112.6	0.0	25.0	50.0	25.0	30.0	34.1	0.0	37.5	50.0	12.5
Lithuania	11	100.0	176.2	0.0	0.0	54.5	45.4	37.0	69.7	0.0	45.4	36.4	45.5

Luxembourg	8	81.7	101.6	25.0	0.0	50.0	0.0	5.2	28.7	62.5	0.0	25.0	25.0
Macedonia	10	96.4	131.5	10.0	20.0	30.0	40.0	53.1	56.6	30.0	10.0	40.0	50.0
Netherlands	13	64.0	60.4	7.7	15.4	76.9	0.0	8.4	8.3	76.9	23.0	0.0	0.0
Norway	12	5.4	12.3	66.7	25.0	8.3	0.0	0.5	4.0	83.3	0.0	8.3	0.0
Poland	67	77.5	106.8	0.0	23.9	55.2	21.0	18.9	33.1	17.9	55.2	22.4	11.9
Portugal	31	2.6	33.0	61.3	12.9	25.8	0.0	1.8	11.5	35.5	19.3	6.4	0.0
Romania	26	113.8	180.3	3.8	23.1	34.6	38.5	38.6	53.3	23.1	23.1	42.3	34.6
Russia	35	74.0	104.9	5.7	31.4	20.0	25.7	30.0	46.0	20.0	22.8	34.3	45.7
Serbia	13	45.7	56.1	30.8	23.1	38.5	7.7	21.9	49.3	23.1	30.8	30.8	7.7
Slovakia	19	213.0	197.9	25.0	21.0	15.8	57.9	73.7	79.5	15.8	10.5	57.9	57.9
Slovenia	15	85.7	145.6	0.0	13.3	60.0	26.7	37.0	232.5	0.0	26.7	46.7	46.7
Spain	130	57.7	71.9	13.1	31.5	50.0	5.4	11.1	24.3	46.9	27.7	17.7	7.7
Sweden	11	55.0	36.4	36.4	9.1	54.5	0.0	5.0	6.3	81.8	18.2	0.0	0.0
Switzerland	54	105.0	188.8	7.4	5.5	53.7	33.3	24.0	29.3	22.2	38.9	37.0	14.8
Turkey	40	21.2	66.3	17.5	52.5	12.5	17.5	5.3	36.2	52.5	15.0	12.5	22.5
Ukraine	19	96.0	122.0	0.0	21.0	52.6	26.3	36.5	50.5	0.0	36.8	47.4	36.8
U.K.	70	35.5	49.2	11.4	50.0	37.1	1.4	6.9	11.1	58.6	32.8	7.1	1.4

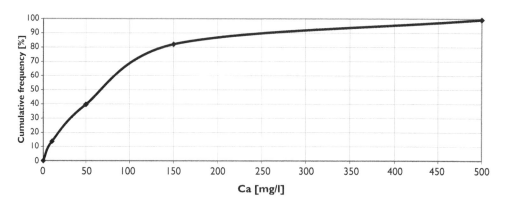

Figure 13.4 Cumulative frequency of Ca content in mineral waters in Europe.

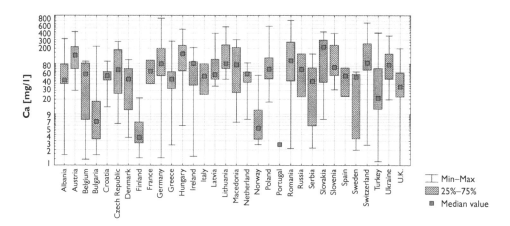

Figure 13.5 Box and Whisker plot of Ca in mineral waters in different countries in Europe.

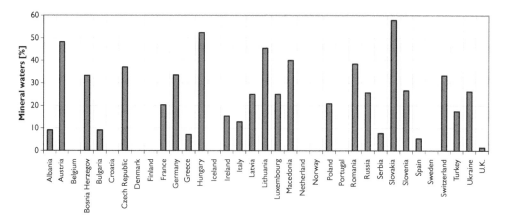

Figure 13.6 Quantity of mineral waters with Ca content exceeding 150 mg/l in European countries.

minerals (Pritchard, 2011). The recommended daily Ca dose for adults ages from 19 to 50 is 1000 mg daily, according to MedlinePlus, a service of the National Institutes of Health. Women over 50 need 1200 mg of calcium daily. Men over 70 also need 1200 mg of Ca daily. The tolerable upper intake level for calcium is between 2000 and 2500 mg daily. Many people, especially those consuming dairy products, have high-calcium diets so they should avoid mineral waters very rich with calcium. Fortunately only 8.8% of European mineral waters has a Ca content higher than 300 mg/l. The highest content of Ca is found in mineral waters from Germany, Italy, Spain, Switzerland, France and Slovakia.

13.3.3   Mg in mineral waters in Europe

The main source of Mg in groundwater is the dissolution of carbonate minerals such as dolomite as well as magmatic minerals such as: olivine, diopside, amphibole and mica during weathering processes. In deep aquifers the reverse ion exchange may be the additional source of Mg in waters as well as dedolomitisation. The most abundant speciations in groundwaters are: Mg^{2+}, $MgSO_4^0$, $MgHCO^{3+}$. Natural background level of Mg in groundwaters is 0.5–50 mg/l (Witczak *et al.*, 2013).

Mg concentration in the European bottled mineral waters that were examined ranges from 0.02 to 1060 mg/l. The average value is 34 mg/l, and the median is 18 mg/l. 33% of European mineral waters show low magnesium content (10–30 mg/l). 37% of waters have even very low Mg content, below 10 mg/l (Figure 13.7). According to EU Directive 80/777/EEC only 21% may be called as 'mineral waters containing magnesium' with Mg concentration above 50 mg/l.

Waters brands with the preferable content of magnesium occur in countries of central Europe (Figure 13.8): Austria, Czech Republic, Hungary, Lithuania, Slovakia, Slovenia and Switzerland. The lowest values of Mg are observed in waters from: Iceland, Finland, Norway, Sweden, Denmark, Albania and the Netherlands.

The optimum content of Mg in bottled mineral and spring waters is more than 50 mg/l. Only in 10 European countries do more than 30% of bottled mineral waters have such high calcium content (Figure 13.9). These are: Slovakia, Macedonia,

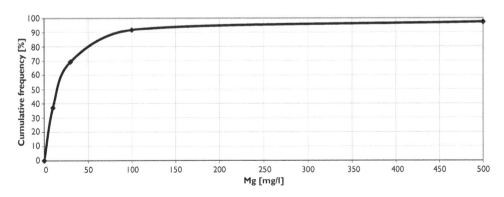

Figure 13.7 Cumulative frequency of Mg content in mineral waters in Europe.

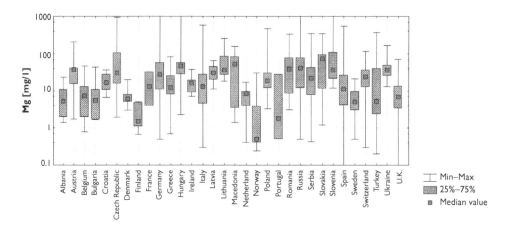

Figure 13.8 Box and Whisker plot of Mg in mineral waters in different countries in Europe.

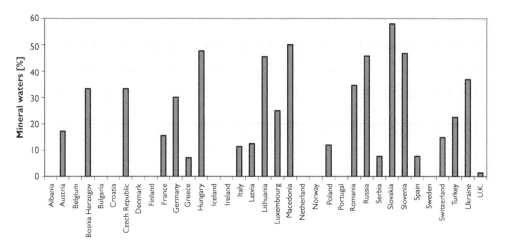

Figure 13.9 Quantity of mineral waters with Mg content exceeding 50 mg/l in European countries.

Lithuania, Hungary, Slovenia, Ukraine, Romania, Czech Republic, Bosnia and Herzegovina and Germany.

In some European countries none of the bottled mineral waters has Mg higher than 50 mg/l. These are 12 countries: Albania, Belgium, Bulgaria, Croatia, Denmark, Finland, Iceland, Ireland, Netherlands, Norway, Portugal and Sweden.

Similar to Ca overdose, Mg overdose may cause some negative effects. A Mg overdose severely lowers blood pressure, slows heart rate and increases the negative side effects of blood pressure medication. Kidney problems have also been associated with excessive Mg levels (Pritchard, 2011). The recommended daily dose of Mg varies according to age and gender. Adults should not take more than 350 mg of Mg daily. Since only 5.7% of European mineral waters have Mg content higher than 100 mg/l there is a low risk of overdosing of Mg from drinking mineral waters.

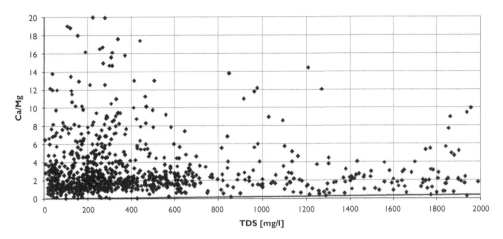

Figure 13.10 The relationship between Ca/Mg ratio and the TDS values of European mineral waters.

The highest content of magnesium has been found in mineral waters from: Slovenia, Czech Republic, Germany, Italy and Slovakia.

13.3.4 Ca/Mg ratio in mineral waters in Europe

For human health the optimal Ca/Mg molar ratio in mineral waters is 2:1. This ratio in mineral waters in Europe ranges from 0.02 to 1102; the average is 4.8 and the median is 2.3. The Ca/Mg ratio is slightly lower in highly mineralised waters and higher in low mineralised waters (Figure 13.10). In mineral waters with the TDS values ranging between 1000 and 1500 mg/l the Ca/Mg median is 2.1; with the TDS ranging between 500 and 1000 mg/l it is 2.2; and for waters with the TDS values higher than 1500 mg/l is 1.8.

In Italy, Portugal, Romania and Sweden less than 20% of waters has Ca/Mg ratio around 2, while in the Netherlands none of the water has. A much better situation occurs in Croatia, Slovenia and Ukraine where more than 40% of waters has Ca/Mg ratio around 2.

13.4 CONCLUSIONS

- The majority of bottled mineral and spring waters in Europe has low a Ca (75%) and Mg (79%) content.
- According to EU Directive 80/777/EEC as little as 18% of European bottled waters may be called 'mineral waters containing calcium' with Ca content higher than 150 mg/l and only 21% of European bottled waters may be called 'mineral waters containing magnesium' with Mg content higher than 50 mg/l.
- This is an alarming situation as the most important nutrient constituents of waters in 75% of bottled mineral and spring waters in Europe are observed in extremely low concentrations.

- Most European countries have no numerical guidelines for Ca and Mg content in bottled drinking water.
- Recommended daily intakes of Ca and Mg should be set at national as well as international levels.

REFERENCES

Bruvold W.H. & Ongerth H.J. (1969) Taste quality of mineralized water. *Amerian Water Works Association*, 61, 170.

Davis W. (2007) Is Your Bottled Water Killing You? *Life Extension Magazine.*

Dissanayake C.B. & Chandrajith R. (2009) *Introduction to Medical Geology*, Erlangen Earth Conference Series, Springer-Verlag Berlin Heidelberg.

Hem J.D. (1989) *Study and Interpretation of Chemical Characteristics of Natural Water.* U.S.G.S. Water – Supply Paper 2254. United States Government Printing Office.pp. 263.

Pritchard J. (2011) http://www.livestrong.com/article/527459-are-magnesium-and-calcium-dangerous.

Reimann C. & Birke M. (Editors) (2010) *Geochemistry of European Bottled Water.* Borntraeger Science Publishers. Germany.

Van der Aa M. (2003) Classification of mineral water types and comparison with drinking water Standards. *Environmental Geology* 44, 554–563.

WHO (2004) Nutrient Minerals in Drinking Water and the Potential Health Consequences of Long-Term Consumption of Demineralized and Remineralized and Altered Mineral Content Drinking Waters, WHO/SDE/WSH/04.01.

WHO (2008). *Guidelines for drinking water quality.* Geneva WHO 2008. http://www.lentech.com/WHO-drinking-waterstandards06.htm. Accessed 9/01/2008.

Witczak S., Kania J. & Kmiecik E. (2013) *Katalog wybranych fizycznych i chemicznych wska źników zanieczyszczeń wód podziemnych i metod ich oznaczania.* (Catalogue of selected physical and chemical groundwater compounds and methods of their detection). Biblioteka Monitoringu Srodowiska. Warsaw.

Wurts W.A. & Masser M.P. (2004) Water hardness-Calcium and Magnesium. Southern Region Aquaculture Center Publication No. 400.

Chapter 14

Ca and Mg in selected medicinal waters of Lower Silesia

Barbara Kiełczawa
Mining Institute, Wroclaw University of Technology, Poland

ABSTRACT

This chapter presents the results of a study of variability of Ca and Mg concentrations in medicinal waters in selected spa resorts in south-western Poland. The highest concentrations of Ca ions are in P-300 intake in Polanica-Zdrój (437 mg/l), Sarenka in Jeleniów (359 mg/l), Pieniawa Chopina in Duszniki-Zdrój (350 mg/l) and Moniuszko (348 mg/l) in Kudowa-Zdrój. The highest Mg ion concentrations (approx. 100 mg/l) occur in waters from the well 2P (102 mg/l) in Świeradów-Zdrój and in Dąbrówka intake in Szczawno-Zdrój. Usually an increase in Ca reflects a parallel increase of Mg. In a few sites (K-200 in Kudowa-Zdrój, P-300 in Polanica-Zdrój, no 5 and 6 in Gorzanow and no 39 and B-2 in Duszniki-Zdrój) an increase in Ca^{+2} ion has the opposite effect with a decrease in the Mg^{+2} ion concentration. Medicinal waters from Szczawno-Zdrój and Świeradów-Zdrój are characterised by small variations in the Ca/Mg ratio.

14.1 INTRODUCTION

The Sudety Mountains belong to a tectonic unit with an abundance of groundwaters meeting the standards for medicinal waters. Most of them have been used in balneology for many years. Hydrochemically, these are mostly naturally carbonated waters, often rich in specific components (Fe, F, Rn). The best known are carbonated waters from health resorts in Kłodzko Land (Polanica-Zdrój, Duszniki-Zdrój, Kudowa-Zdrój) and the areas near Wałbrzych (Szczawno-Zdrój, Jedlina) and Świeradów (Świeradów-Zdrój, Czerniawa). Sudetic thermal waters are classified as medicinal waters.

The procedures related to monitoring and supervision of medicinal waters intakes in each health resort include monitoring and analyses of physico-chemical properties of waters from particular intakes. The results of these analyses are used here to investigate concentrations of calcium (Ca^{+2}) and magnesium (Mg^{+2}) ions in selected waters. The choice was based on the accessibility of essential physico-chemical analyses. The results of chemical composition analyses were gathered for waters from particular intakes located in Kudowa-Zdrój, Jeleniów, Duszniki-Zdrój, Polanica-Zdrój and Gorzanów, as well as Szczawno-Zdrój and Świeradów-Zdrój.

The Sudetes as a tectonic unit form the north-eastern part of the margin zone of the Bohemian Massif, the largest crystalline massif in Central Europe (Oberc, 1972). Many different tectonic processes that the Sudetes have undergone in their development have produced a mosaic-like structure (Figure 14.1). Consequently, in a

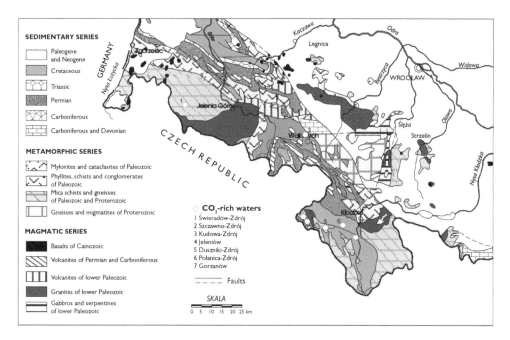

Figure 14.1 Distribution of sampled waters and geological map of the Sudety (based on Sawicki, 1995).

relatively small area, outcrops of crystalline, volcanic and metamorphic rocks occur alongside sedimentary series (Przylibski, 2005).

The medicinal waters of Szczawno-Zdrój are found within Lower-Carboniferous deposits. These are alternating beds of conglomerates, greywacke sandstones and mudstones – red in colour due to the presence of haematite accumulated in grey-wackes and conglomerate cement (Ciężkowski *et al.*, 1996f). North east of Szczawno there are outcrops of Pre-Cambrian gneisses in the Sowie Mts.

In the Świeradów area are mostly Pre-Cambrian crystalline formations composed of different varieties of gneiss and schists (Ciężkowski *et al.*, 1996g). These rocks form a large tectonic unit – the Izera Metamorphic Complex, shielding the granite Karkon-osze Massif from the north.

Waters of Kudowa-Zdrój and Jeleniów occur in a small geologic unit known as the Kudowa depression. This structure, whose basement is built of Early-Palaeozoic metamorphic series (mica, chlorite-biotite and quartzite schists as well as sericite phyl-lites), is filled with the Carboniferous and Permian (alternating layers of conglomer-ates, arcose sandstones, clay shales, sand shales and conglomerates) and Cretaceous (lime conglomerates, sandstones, mud shales as well as clayey-siliceous and siliceous marls) sediments. The whole structure forms a synclinal depression surrounded by granitoids (Ciężkowski, 1990; Ciężkowski *et al.*, 1996c–d).

The area with the Duszniki-Zdrój waters comprises metamorphic and sedimentary rocks, varying in age (Figure 14.1). The oldest are Palaeozoic quartzite, mica, and graphite schists, gneisses and crystalline limestones. They contain Late-Palaeozoic porphyry intrusions. This metamorphic complex is covered with Upper-Cretaceous sediments filling the Duszniki depression. They are generally interbedded conglomerates, sandstones, mudstones, as well as sand and clayey sand marls. Cretaceous sedimentation in the Duszniki area is complete with clayey-siliceous or sand marls and quartz sandstones (Ciężkowski *et al.*, 1996a).

Mesozoic sedimentary series and their crystalline basement occur in the discharge areas of Polanica-Zdrój and Gorzanów. The metamorphic basement is built of gneisses and biotite-muscovite-chlorite mica schists with quartz insertions. North west of Polanica the Permian sediments are covered directly by Upper-Cretaceous (Cenomanian, Turonian and Coniacian) deposits. The Cenomanian deposits were formed mostly in sand facies. In the bottom section, these are grey clayey-limy sandstones underlain with quartz conglomerates. The next level is formed of quartz sandstones, so-called lower jointed sandstones. The Upper Cenomanian is formed of marly sandstones (Radwański, 1975). They are overlain by Turonian sediments, which are predominantly clayey and clayey-siliceous marls, interbedded with jointed sandstones and insertions of sand marls, claystones and sandy limestones. The Coniacian formation is dominated by deposits that developed in a flysch facies i.e. sandstones, mudstones and claystones occurring in an alternating sequence (Ciężkowski *et al.*, 1996b, e).

14.2   RESULTS AND DISCUSSION

The discharge of the waters are predominantly located near tectonic zones or at intersections of principal faults (e.g. Struga-Szczawno dislocation, or Hronov-Gorzanów – Gorzanów zone) and secondary dislocations (e.g. Duszniki source fault, Polanica-Wolany fault-Polanica) (Ciężkowski, 1990).

The waters are carbonated waters slightly varying in hydrochemical types:

HCO_3–Ca and HCO_3–Ca (–Na) – Polanica, Gorzanów,
HCO_3–Ca–Mg – Duszniki, Świeradów,
HCO_3–Ca–Na (–Mg) – Duszniki,
HCO_3–Mg–Ca – Świeradów,
HCO_3–Na–Ca – Kudowa, Jeleniów, Szczawno.

The concentrations of Ca^{+2}, Mg^{+2}, Na^+ ions in some waters are higher than 20%mval.

The highest concentrations of Ca^{+2} ions characterise waters from intakes: P-300 in Polanica-Zdrój (437 mg/l), Sarenka in Jeleniów (359 mg/l), Pieniawa Chopina in Duszniki-Zdrój (350 mg/l) and Moniuszko (348 mg/l) in Kudowa-Zdrój. The lowest concentrations (below 50 mgCa/l) are found in waters from Świeradów-Zdrój (apart from borehole 2P), Szczawno-Zdrój (less than 100 mg/l) and intake Kaczka (68 mg/l) in Gorzanów (Table 14.1, Figure 14.2).

Variations in the concentration of Ca^{+2} depend on the TDS of the waters, an increase in the Ca^{+2} ions accompanies an increase in TDS (Figure 14.3). Such a relation suggests mixing of high TDS deep circulation waters with low TDS shallow circulation waters (Ciężkowski, 1990; Kiełczawa, 2001). The variation in the number of Ca^{+2} ions could also result from ion exchange or from precipitation of secondary calcite in rock fissures (Kiełczawa, 2011).

Table 14.1 Ca^{+2} and Mg^{+2} ion concentrations (in mg/l) in waters from particular medicinal water intakes in selected spa resorts.

Spa resort	Medicinal water intake	Ca^{+2} [mg/l]			Mg^{+2} [mg/l]			$\dfrac{Ca^{+2}}{Mg^{+2}}$
		Min	śr	Max	Min	śr	Max	
Polanica-Zdrój	P-300	63.3	367.3	436.9	16.5	58.8	94.5	6.25
	Józef-2	120.4	138.9	153.3	11.2	16.5	19.8	8.42
	Józef-1	148.2	184	273.6	14.3	21.1	33.2	8.72
	Józef Stary	183.2	191.3	208.4	22.8	26.9	30	7.11
	Wielka Pieniawa	226.7	261.5	316.4	17.8	30.6	39.4	8.55
	Żelaziste	244	264.7	292.6	12.2	25.2	30.8	10.50
Gorzanów	Kaczka	68.6	86.9	95.3	7.6	11.9	14.7	7.30
	nr 5 gosp.	108.5	163.4	206	17.5	21.6	29.1	7.56
	7M	144	154.2	168.2	17.3	21	24.3	7.34
	nr 6	162	186.4	205.6	15.2	22.8	28.2	8.18
	nr 4b	182.5	190.8	200	18.9	21.2	23.5	9.00
	nr 5 prod.	263.5	277.8	289	19	22.5	31	12.35
Kudowa-Zdrój	Marchlewski	128	180.7	224.4	19	32.7	39.4	5.53
	Górne	154.7	172.6	189.6	31.3	46.6	57.8	3.70
	Śniadecki	163	200.8	221.6	49	58.2	64.12	3.45
	K-200	216.6	225.3	238.9	73.1	84.5	92.5	2.67
	Moniuszko	290	327.9	348.1	36	67.1	88	4.89
Jeleniów	J-150	87.3	104.2	137.1	26.1	34.7	43.9	7.02
	Sarenka	101	247	358.6	16.5	35.2	53	3.00
Duszniki-Zdrój	B-4	70.5	256	292.6	24.2	80.4	94	3.18
	Zimny Zdrój	110.7	127.5	154.7	24	33.2	37.3	3.84
	B-2	111.8	123.2	130	65	67.8	71.6	1.82
	Jan Kazimierz	125	158.7	186.5	25.8	40.5	57.6	3.92
	Agata	143.7	248.6	314.6	47.1	69	86.7	3.60
	B-3	160	214.5	296	57.9	81	98	2.65
	nr 39	193.5	208.7	222	52.1	59.6	65.1	3.50
	B-1	194.2	239	291.2	48.7	71	87.8	3.37
	Pieniawa Chopina	208	233	350.3	51.3	64	79.9	3.64
Szczawno-Zdrój	Marta	83.4	109.6	195.4	32.2	52.6	99.8	1.64
	Mieszko	85.9	107.4	127	49.7	65.5	77	1.64
	Dąbrówka	88.4	112.1	156.9	12.2	73.5	87.6	1.65
	Młynarz	89.8	105.3	133.8	52.4	63.8	82	2.08
Świeradów-Zdrój	Górne-A	12.5	27.08	59	4.77	10.22	23	1.53
	Górne-B	16.5	28.42	50.41	5.1	10.04	19.24	2.65
	Górne-zb.	23.5	38.59	62.1	9.7	18.31	31.6	2.83
	Odwiert-1A	45	53.40	73.75	29.55	37.99	47.6	2.11
	Odwiert-2P	93.2	122.92	190.43	72.9	86.06	102.3	1.41

The largest concentrations of Mg^{+2} ions (about 100 mg/l) were found in waters from borehole 2P (c.102 mg/l) in Świeradów-Zdrój and intake Dąbrówka in Szczawno-Zdrój (Table 14.2, Figure 14.4). Slightly lower concentrations of this cation were observed in the waters in Duszniki-Zdrój (intakes B-3, B-4, B-1 and Agata successively), Kudowa-Zdrój (K-200, Moniuszko) and intake P-300 in Polanica-Zdrój

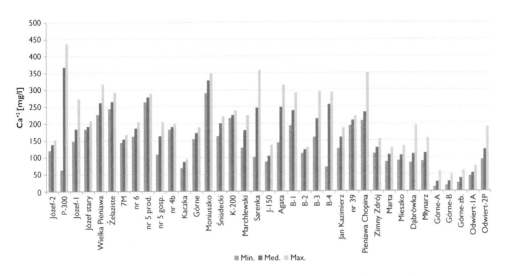

Figure 14.2 Ca^{+2} ions concentrations (mg/l) in selected intakes of medicinal waters.

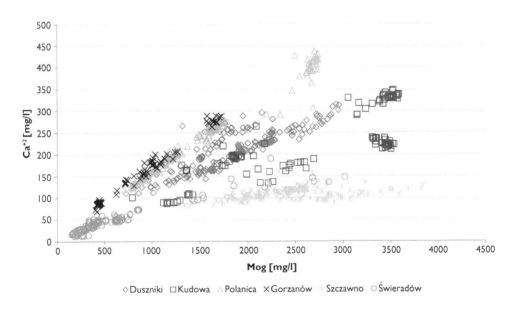

Figure 14.3 Variation in Ca^{+2} ion concentration (mg/l) with TDS of the waters.

(Table 14.2). Also an increase in Mg concentration corresponds to a raise in the TDS values (Figure 14.5).

A simultaneous change in the Ca and Mg concentrations takes place in waters from most intakes (Figure 14.6a). Only in intakes K-200 (Kudowa-Zdrój), P-300 (Polanica-Zdrój), No 5-prod. and No 6 (Gorzanów), as well as No 39 and B-2

Table 14.2 Ca^{+2} and Mg^{+2} ion concentrations (%mval) in waters from selected medicinal water intakes in spa resorts.

Spa resort	Medicinal water intake	Ca^{+2} [%mval]			Mg^{+2} [%mval]		
		Min	Max	Average	Min	Max	Average
Polanica-Zdrój	P-300	50.1	83.8	63.4	12.3	32.9	17.3
	Józef-2	53.8	67.5	63.9	8.9	14.3	12.5
	Józef-1	59.9	85	65.8	7.75	17.9	12.4
	Józef Stary	63	81.6	67.1	13.4	16.9	15.5
	Wielka Pieniawa	25.7	86	65.2	5.2	16.6	12.6
	Żelaziste	62.7	87.9	68.3	6.0	12.9	10.6
Gorzanów	Kaczka	70.6	83.4	77.2	11	21.2	17.4
	no 5-gosp.	61.6	69.5	66.1	11.4	23	15.1
	no 2	73.8	83.6	76	12.2	13.8	12.9
	no 6	61.5	69.7	66.4	8.1	16.5	13.4
	no 4b	72.5	75.9	73.4	11.8	15.1	13.6
	no 5-prod.	64.1	73.1	67.4	7.4	11.9	9
Kudowa-Zdrój	Marchlewski	38.6	48.1	42.5	9.9	16.6	12.7
	Górne	26.2	31.9	29.3	10.65	14.8	12.9
	Śniadecki	26.4	69	58.5	12.5	32.4	27.4
	K-200	24.7	30.3	26.77	14.04	17.7	16.5
	Moniuszko	34.4	43.1	38.3	7.9	17.2	12.9
Jeleniów	J-150	29.2	36.1	32.1	15.4	19	17.6
	Sarenka	48.4	70	55.9	12.4	18.1	14.8
Duszniki-Zdrój	B-4	33.5	45.6	41.1	19.24	23.2	21.4
	Zimny Zdrój	52.3	61.7	56.8	22.1	28.9	24.4
	B-2	43.8	50.7	48.5	42.8	46.2	44.0
	Jan Kazimierz	39.7	70.1	44.9	13.7	27.0	18.8
	Agata	38.6	67.8	44.7	18.5	29.8	20.5
	B-3	48.4	55.1	52.3	28.0	37.0	32.8
	no 39	3.0	49.4	45.3	19.9	23.0	21.6
	B-1	40.6	46.3	43.2	18.8	24.5	21
	Pieniawa Chopina	41.1	67.9	44.6	16.4	29.2	20.2
Szczawno-Zdrój	Marta	16.5	21.1	18.9	16	22	19
	Mieszko	11.9	16.7	14	9.5	17.3	14
	Dąbrówka	18.2	26.8	21.9	14.3	21.4	17.2
	Młynarz	17.1	23.1	19.6	4	23.6	21.2
Świeradów-Zdrój	Górne-A	19.1	59.9	38.4	18.1	39.5	26.3
	Górne-B	39.9	54.4	47.6	15.6	38	26.8
	Górne-zb.	35.5	63.7	42.2	25.3	39.8	32.3
	well-1A	28.7	36.3	33.2	31.9	45	39.1
	well-2P	20.2	34.2	26.6	21.4	40.4	31.6

(Duszniki-Zdrój), does a growth in Ca^{+2} concentration occur alongside a decrease in the Mg^{+2} ion concentration (Figure 14.6b).

One of the methods of defining concentrations of particular ions is expressing their content in %mval so that the chemical type of the water may be determined. Thus, waters with the highest (from c. 70%mval to the maximum of 88%mval) relative con-

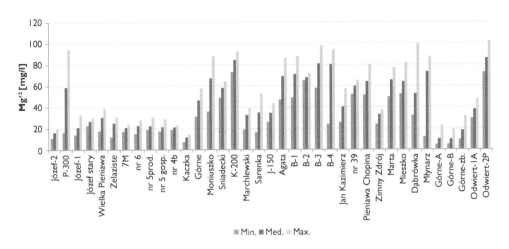

Figure 14.4 Mg⁺² ion concentrations (mg/l) in selected medicinal water intakes.

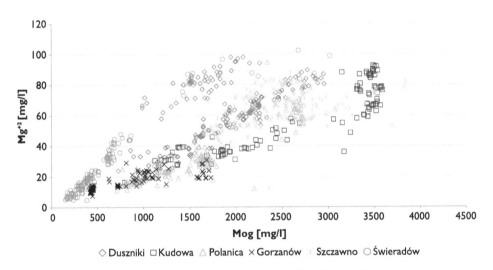

Figure 14.5 Variation in Mg⁺² ion concentration (mg/l) with the TDS of the waters.

tent of Ca^{+2} ions are extracted in Polanica-Zdrój and Gorzanów (Figure 14.7a). Waters from the remaining spa resorts show a slightly lower proportion Ca^{2+} (c. 60–65%mval at the most), but they are all dominated by hydrogen carbonate and calcium ions (HCO_3–Ca).

Waters from Szczawno-Zdrój contain less than 20%mval of Ca^{+2} ions (Figure 14.7a) with Mg^{+2} ion concentrations slightly higher than 20%mval (Figure 14.7b).

HCO_3–Ca–Mg waters are found in Duszniki-Zdrój and Świeradów-Zdrój. The highest relative contents of Mg^{+2} ions (Table 14.2) are observed in intakes No 39,

a)

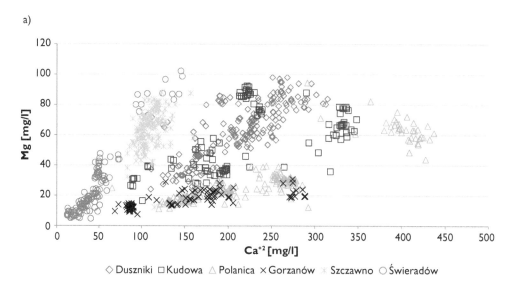

◇ Dusznki   □ Kudowa   △ Polanica   ✕ Gorzanów   ✳ Szczawno   ○ Świeradów

b)

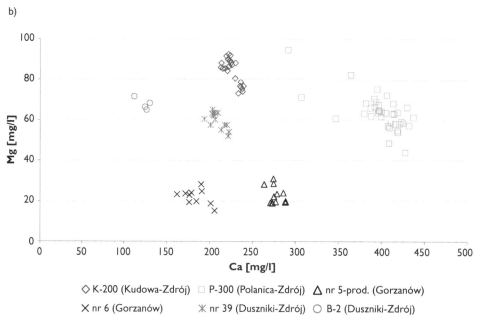

◇ K-200 (Kudowa-Zdrój)   □ P-300 (Polanica-Zdrój)   △ nr 5-prod. (Gorzanów)

✕ nr 6 (Gorzanów)   ✳ nr 39 (Duszniki-Zdrój)   ○ B-2 (Duszniki-Zdrój)

Figure 14.6 Variation in Mg^{+2} ion concentration with Ca^{+2} ion content (mg/l): a) in waters from selected health resorts, b) in waters from selected intakes.

B-2, B-3 (Duszniki-Zdrój) and Górne zb., borehole 1A and borehole 2P (Świeradów-Zdrój). Both the TDS of waters and the concentrations of particular ions are variable. Their variations in time may be so large that the chemical types of waters can change periodically. Occasionally, the concentrations of Mg^{+2} ions are so significant (above 20%mval) that some waters can be regarded as HCO_3–Ca–Mg type (Table 14.2). Such a phenomenon can be observed e.g. in Polanica-Zdrój (intake P-300), Szczawno-Zdrój (Marta and Dąbrówka), Duszniki-Zdrój (Jan Kazimierz) or Gorzanów (Kaczka, No 5-prod.).

The Ca/Mg ratios vary from 1.4 (borehole 2p in Świeradów-Zdrój) to the maximum of 12.4 (No 5-prod. in Gorzanów). Generally, the waters of Szczawno-Zdrój and Świeradów-Zdrój are characterised by the smallest variations in this ratio (Table 14.1). The predominance of Ca^{+2} ion over Mg^{+2} (from c. 6 to more than 12) is characteristic of waters from Polanica-Zdrój and Gorzanów.

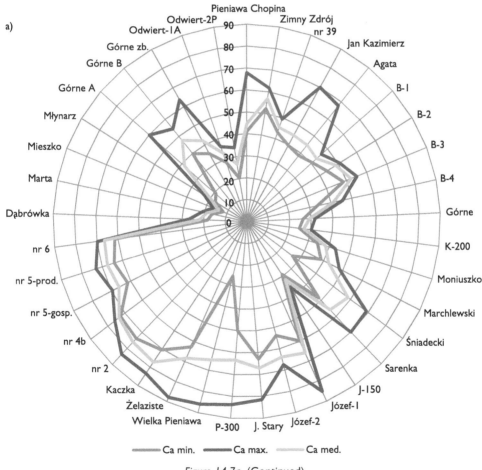

Figure 14.7a (Continued).

b)

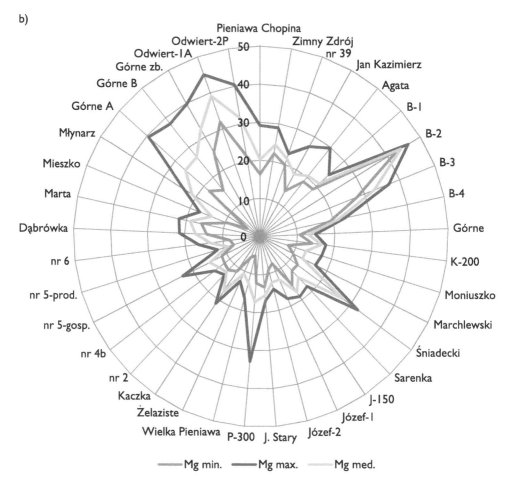

Figure 14.7b Ionic concentrations (%mval/l) in selected medicinal water intakes: a) Ca^{+2} ions, b) Mg^{+2} ions.

14.3　CONCLUSIONS

The chemical composition of the carbonated waters is determined mainly by their reactions with the rock and the carbon dioxide (of deep origin) migrating towards the intake areas. The largest influence is exerted by dissolution of primary silicates, aluminosilicates and crystalline limestones, and crystallisation of secondary carbonate (calcite) and clayey (kaolinite) minerals. In addition, calcium ions can be replaced with Na^+ ions from clay minerals.

ACKNOWLEDGEMENT

The work was carried out in the framework of statute research No S 10026.

REFERENCES

Ciężkowski W. (1990) Studium hydrogeochemii wód leczniczych Sudetów polskich (Hydrogeochemistry of medicinal waters in polish Sudetes). *Pr. Nauk. Inst. Geotech. PWroc.*, 60, Wrocław.

Ciężkowski W., Doktór St., Graniczny M., Izydorska A., Kabat T., Liber-Madziarz E., Przylibski T., Teisseyre B., Wiśniewska M., Zuber A. (1996) Próba *określenia obszarów zasilania wód leczniczych pochodzenia infiltracyjnego w Polsce na podstawie badań izotopowych (Delineation of recharge areas of medicinal waters in Poland basing on isotopic data)*. Appendices: a – medicinal waters of Dusznik-Zdrój, b – medicinal waters of Gorzanów, c – medicinal waters of Jeleniów, d – medicinal waters of Kudowa-Zdrój, e – medicinal waters of Polanica-Zdrój, f – medicinal waters of Szczawno-Zdrój, g – medicinal waters of Świeradow-Zdrój.

Kiełczawa B. (2001) Zjawisko mieszania się wód Gorzanowa na tle złóż wód rowu górnej Nysy Kłodzkiej. (Mixing of waters in Gorzanow on the background of resources of Gorna Nysa Kłodzka trough), *Pr. Nauk. Inst. Gór. PWroc.* 92(29), 85–94.

Kiełczawa B. (2011) The main hydrogeochemical processes affecting the composition of certain naturally carbonated waters of southwestern Poland, *Geol. Quart.* 55(3), 203–212.

Oberc J. (1972) Budowa geologiczna Polski. Tektonika, Sudety i obszary przyległe, (Geology of Poland. Tectonics. Sudetes and surroundings) Wyd. Geol. W-wa. 4(2), p. 307.

Przylibski T.A. (2005) *Radon, składnik swoisty wód leczniczych Sudetów (Radon, specific komponent of medical waters of Sudetes)*, Oficyna Wyd. PWroc., Wrocław, p. 329.

Radwański S. (1975) Kreda Sudetów środkowych w świetle wyników nowych otworów wiertniczych. (Cretacerous of Middle Sudetes in the results of investigations of new boreholes), *Biul. Inst. Geol.*, 24(28), 5–50.

Sawicki L. (1995) Mapa geologiczna regionu dolnośląskiego z przyległymi obszarami Czech i Niemiec (bez utworów czwartorzędowych) 1:100 000, Państw. Inst. Geol., Warszawa.

Chapter 15

Type and concentration of different compounds of Ca and Mg in natural waters used in therapeutic treatment or as food (bottled waters)

Danuta Sziwa, Teresa Latour & Michał Drobnik
National Institute of Public Health – NIH, Department of Health
Resort Materials, Poznań, Poland

ABSTRACT

Use groundwater for human needs, such as health, consumption and hygiene, is determined by the content of calcium (Ca) and magnesium (Mg) concentrations, and co-existence of macro-components such as sodium (Na^+), chloride (Cl^-), sulfate (SO_4^{2-}), bicarbonate (HCO_3^-) and carbon dioxide (CO_2). Groundwaters occurring in Poland are very diverse. In recent years the National Institute of Public Health – National Institute of Hygiene has carried out a large number of analyses of chemical and physico-chemical parameters on a wide range of groundwater types. The results of these analyses indicate that the concentration of Ca^{2+} and Mg^{2+} ions is highest (up to several grams/l) in some of the brines used for medical purposes. Naturally-carbonated water is characterised by a significant amount of both elements (up to several hundred mg/l), while also containing large a concentration of bicarbonate. Another example is alkaline water, which contains very low amounts of calcium (just afew mg/l) and magnesium (less than 1 mg/l), a significant concentration of sodium bicarbonate and has pH > 8. There are various Ca^{2+} to Mg^{2+} ionic ratios in Polish groundwaters, although the amount of calcium is usually at least two times higher than magnesium. In Poland rare types of waters are magnesium-sulfate waters and waters with a dominance of Mg^{2+} ion. The different types of Polish groundwaters rich in Ca and Mg are described in this chapter. Some groundwaters can possibly be used in medicine, by the food industry and for cosmetics.

15.1 INTRODUCTION

The calcium (Ca) and magnesium (Mg) concentrations in therapeutic, bottled and drinking water are important. Ca and Mg are the so-called macronutrients. In many groundwater, particularly from shallow aquifers, their concentration and coexistence with other ions: bicarbonate, sulfate, chloride, carbon dioxide, determine the use of the water. The characteristic water composition and its Total Dissolved Solids (TDS) determine the possibility of using this water in different areas of human ativity: hydrotherapy, inhalation, crenotherapy (water cure) and the daily consumption of water. Each type of water (therapeutic water, bottled water, water to drink) has its own regulations described in the *Journal of Law, Poland*. Various waters have been analysed in this study along with their role as a significant source of Ca and Mg in human diet.

15.2   STUDY AREA AND RESULTS

The results derive from a long-term programme of research and analysis of water for spa treatments and the suitability of water for the production of bottled water (natural mineral water, spring water and table water according to European Union regulations). Taking into account the degree of absorption of Ca and Mg in the body, and the impact of other constituents on natural absorption of Ca and Mg, a significant physiological effect is achieved if the natural water contains at least 500 mg/l Ca and at least 150 mg/l Mg and acquire the status of therapeutic waters, with indications or contraindications to their use.

15.2.1   The therapeutic water

Therapeutic (curative) water is mineral water and/or specific water, distinguished by its stable specific physical and chemical characteristics (within the allowable fluctuation), meeting the sanitary and therapeutic requirements set by the Council of Ministers, and meeting at least one of the following conditions:

- the total dissolved mineral solids >1000mg/l,
- ferrous ion content >10 mg/l (ferruginous water),
- fluoride ion content >2 mg/l (fluoride water),
- iodide ion content >1 mg/l (iodide water),
- sulfur(II) content >1 mg/l (sulfurous water),
- metasilicic acid content >70 mg/l (Si-rich/siliceous water),
- radon content >74 Bq (radon/radioactive water),
- carbon dioxide free >250 mg/l (250–999 mg/l = CO_2-rich water, carbon dioxide ≥1000 mg/l = acidulous water),
- temperature >20 °C at water intake (thermal water).

15.2.2   Calcium and magnesium in various types of therapeutic waters in Polish health-resorts

The highest concentrations of Ca (4.0–8.0 g/l) and Mg (1.56–2.4 g/l) were found in brines from Ustroń and Goczałkowice (Table 15.1). These waters are mineralised from 8.0 to 10.0%, and have been used for bathing and for inhalation. The lowest concentrations of these components especially Mg (<1 mg/l), were found in the waters at Lądek Zdrój and Cieplice Śląskie Zdrój (alkaline pH), but also in some radon (Rn) waters from Świeradów-Czerniawa. In most cases Ca dominates the Ca: Mg ratio in the therapeutic waters. Sometimes in waters of HCO_3–Cl–Na type, J, CO_2 type from Iwonicz Zdrój and Szczawnica and the Zuber's (from Krynica Zdrój), and the Z-II (750 mg Mg, 190 mg Ca) the situation changes with the Ca: Mg ratio at 2:1 or 3:1, ocasionally 10:1.

The waters were determined to be suitable for curative purposes (drinking treatment) for various reasons, in addition to the Ca and Mg content. The maximum concentration of Ca is about 400 mg/l, and Mg also about 400 mg/l. Much higher concentrations of these elements occur in the brines (Ca up to 8000 mg/l, Mg 2480 mg/l in Ustroń) where they play a physiological role for bathing, inhalation and rinsing body cavities.

Table 15.1 Ca and Mg in various types of therapeutic waters in Polish health-resorts. Medical treatment: Inhalations and water cure.

Health-resort	Number of water intakes	Mineralisation mg/l	Type of water	Content in mg/l		Medical treatment
				Ca	Mg	
The chemical type: brines & saline waters						
Ciechocinek	7	0.33–6.8%	Cl–Na, J, T	138.0–2068.0	58.0–667.0	inhalations
Kołobrzeg	3	0.46–5.53%	Cl–Na, J	183.0–1866.0	72.0–650.7	inhalations
Konstancin	1	7.5%	Cl–Na, J	2156.0	685.0	inhalations
Goczałkowice	2	6.8–7.8%	Cl–Na, J	3360.0–4100.0	1300.0–1560.0	inhalations
Busko-Zdrój	5	1.3–2.3%	Cl–Na, J, S	304.0–570.0	237.0–380.0	water cure
Solec	3	1.6–2.0%	Cl–Na, J, S	500.0–921.0	346.0–425.0	inhalations
Rabka	5	1.7–2.2%	Cl–Na, +HCO$_3^-$, J	40.0–105.0	27.0–49.0	inhalations
Ustroń Śląski	2	10.0%	Cl–Na, J	8000.0–9440.0	average: 2480.0	inhalations
Sopot	1	4.2%	Cl–Na, J	2400.0	1200.0	inhalations water cure prod. of salt
The chemical type: HCO$_3$–Cl–Na, J, CO$_2$						
Iwonicz Zdrój	7	0.6–1.9%	HCO$_3$–Cl–Na, J, CO$_2$	26.0–45.0	8.0–59.0	water cure inhalations prod. of salt
Szczawnica	7	0.2–2.6%	HCO$_3$–Cl–Na, J, CO$_2$	120.0–353.0/110.0	28.0–80.0/241.0	water cure inhalations
The chemical type: HCO$_3$–Ca–Mg/Na, CO$_2$, also J						
Piwniczna	2	1.9%	HCO$_3$–Ca–Mg	263.0–324.0	71.0–133.0	water cure
Rymanów	4	0.93–0.86%	Cl–HCO$_3$–Na, J, CO$_2$	15.0–186.0	14.0–31.0	water cure
Wysowa	8	0.1–2.5%	HCO$_3$–Cl–Na/Ca, CO$_2$	90.0–314.0	22.0–75.0	water cure

(Continued)

Table 15.1 (Continued).

Health-resort	Number of water intakes	Mineralisation mg/l	Type of water	Content in mg/l Ca	Mg	Medical treatment
Krynica Zdrój	20 all water: CO_2–rich	I type: 0.1–1.6%	HCO_3–Ca	200.0–329.0	21.0–51.0	water cure
		II type: 2.5–2.9%	HCO_3–Na	201.0–220.0	378.0–494.0	water cure (Zuber I, III, IV)
		III type: 2.2%	HCO_3–Na–Mg	190.0	750.0	water cure (Zuber II)
		IV type: 012–0.8%	HCO_3–Ca–Mg/Na	196.0–924.0	117.0–636.0	water cure
Żegiestów	3	0.2–1.4%	HCO_3–Ca–Mg and HCO_3–Na–Ca–Mg	283.0–385.0	109–1213.0	water cure
The chemical type: HCO_3–Na or low-mineralisation, specific water, also Rn						
Szczawno	4	0.2–0.3%	HCO_3–Na	95.0–124.0	45.0–82.0	water cure inhalations
Polanica	3	0.1–0.3%	HCO_3–Ca	131.0–394.0	17.0–76.0	water cure inhalations
Długopole	3	0.08–0.13%	HCO_3–Ca–Mg–Na	89.0–130.0	37.0–60.0	inhalations water cure
Kudowa	4	0.13–0.35%	HCO_3–Ca–Na	155.0–332.0	20.0–76.0	inhalations water cure
Świeradów	8	0.006–0.1%	Rn	5.0–104%	<1.0–47.0	inhalations
Czerniawa	6	0.08–0.3%	Rn	10.0–90.0/345.0	3.0–56.0/150.0	inhalations
Cieplice	8	~0.06–0.09%	F, T	10.0–70.0	<1.0–10.0	inhalations water cure

15.2.3   Ca and Mg in bottled waters

According to *The list of recognized water as natural mineral water* there are 87 natural mineral water brands in Poland, divided into three categories:

- The highly-mineralised water–with TDS >1500 mg/l,
- The medium-mineralised water–with TDS from 500 to 1500 mg/l,
- The low-mineralised water–with TDS <500 mg/l.

High concentrations of Ca and Mg are found only in the highly mineralised waters and are rare in the medium-mineralised waters. Usually these waters are of HCO_3–Ca–Mg and HCO_3–Cl–Ca–Mg types, naturally saturated with CO_2 in the spring. To achieve positive health effects rich Ca and Mg mineral water should be consumed systematically, from 1.5 to 2.0 l/day, in divided doses.

The highly-mineralised waters (the most popular brands) have a TDS ranging between 1548 to 3333 mg/l. The Ca content ranges from 86 mg/l to 482 mg/l, and Mg from 26 mg/l to 139 mg/l (Table 15.2).

The medium-mineralised waters (the most popular brands) have a TDS in the range 500 to 1290 mg/l, with Ca content between 62 and 196 mg/l, and Mg content between 13 and 61 mg/l (Table 15.3).

The low-mineralised waters (the most popular brands) are of the TDS ranging from 320 to 498 mg/l, Ca content between 44 and 97 mg/l, and Mg content between 4 and 23 mg/l (Table 15.4).

The spring water (the most popular brands) are of the TDS ranging from 232 to 679 mg/l, Ca content between 5 and 112 mg/l, and Mg content between 2 and 27 mg/l (Table 15.5).

Among the bottled, highly-mineralised waters, the Ca content ranges from 86–482 mg/l and the Mg content between 26 and 139 mg/l (about 15 bottled waters). These are often naturally saturated with CO_2. The chemical type of water is HCO_3–Ca–Mg or HCO_3–Cl–Ca–Mg.

Table 15.2 High-mineralisation water (above 1500 mg/l).

Name	Ca mg/l	Mg mg/l	HCO₃ mg/l	Total mineralisation mg/l
1. Krystynka	174	63	470	3333
2. Kryniczanka	437	63	470	2460
3. Muszyna Minerale	**482**	52	1855	2374
4. Galicjanka	302	69	1705	2211
5. Staropolanka 2000	325	54	1620	2193
6. Zdroje Piwniczna	222	101	1579	2100
7. Wysowianka	86	26	1172	1973
8. Krynica Minerale	344	56	1391	1880
9. Muszynianka	176	**139**	1396	1857
10. Polanica Zdrój	313	34	1281	1790
11. Piwniczanka	172	90	1289	1761
12. Muszyna Zdrój	243	49	1153	1581
13. Muszyna	319	32	1154	1548

Table 15.3 Medium-mineralisation water (500–1500 mg/l).

Name	Ca mg/l	Mg mg/l	HCO₃ mg/l	Total mineralization mg/l
1. Kropla Minerałów	160	61	977	1290
2. Minervita	196	53	500	1243
3. Polanicka	161	15	625	972
4. Staropolanka	149	16	628	895
5. Augustowianka	70	23	414	886
6. Viva Minerale	62	21	417	819
7. Buskowianka	121	30	365	782
8. Cisowianka	116	21	480	669
9. Nałęczowianka	114	20	448	650
10. Nałęczów Zdrój	107	21	461	636
11. Kazimierska	88	29	388	589
12. Polaris	97	18	429	588
13. Wieniecka Zdrój	94	18	305	524
14. Kinga Pienińska	98	13	336	513
15. Jurajska	66	33	330	500

Table 15.4 Low-mineralisation water (<500 mg/l).

Name	Ca mg/l	Mg mg/l	HCO₃ mg/l	Total mineralization mg/l
1. Ustronianka Biała	93	18	305	498
2. Polaris Plus	97	14	254	496
3. Perła Połczyńska	81	11	279	444
4. Arctic Plus	74	13	260	414
5. Jura Skałka	44	23	236	334
6. Kropla Beskidu	46	19	202	324
7. Galicya	59	4	191	320

Table 15.5 Spring water (the most popular brands).

Name	Ca mg/l	Mg mg/l	HCO₃ mg/l	Total mineralization mg/l
1. Aquarel	112	24	482	679
2. Dar Natury	88	19	341	522
3. Ola	86	16	284	470
4. Dobrawa	55	27	284	405
5. Amita	5	13	171	355
6. Złoty Potok	71	5	217	336
7. Mama i Ja	44	5	165	269
8. Aleksandria	46	2	171	264
9. Primavera	50	6	170	256
10. Victoria Cymes	45	7	172	256
11. Żywiec Zdrój	43	6	136	232

In a few (10%) medium-mineralised waters, the Ca content is within the range 150 to 200 mg/l, and the Mg between 50 and 100 mg/l. The ratio of Ca: Mg is 2:1 in the preferred waters. The content of Ca and Mg in the spring waters and low-mineralised waters is at the same level as in the drinking water. The presence of CO_2 and ionized forms of chemical elements and compounds in the water are factors which help in the absorption of Ca and Mg.

15.2.4   Ca and Mg in drinking water

According to the current regulations there are no specific limits for calcium, magnesium and hardness in waters for a human consumption within Europe. The drinking water is usually abstracted from shallow aquifers. This water has total mineralisation in the range 200 to 400 mg/l. Typically, Ca as bicarbonate dominates and Ca, Mg, Na as bicarbonate and sulfate prevail.

Ca and Mg in drinking waters from water public supplies in the main cities in Poland are diverse (Table 15.6).

The highest concentrations of a are in Gdańsk, Bydgoszcz and Zielona Góra, and the highest concentrations of Mg are in Poznań and Sopot. Drinking water may be an additional source of Ca and Mg, because they are essential in our daily diet (Tables 15.7 and 15.8). Ca and Mg supplementation occurs in some drinking waters (Zachwieja, 2006).

15.2.5   Assimilation of Ca & Mg from water. Dietary Reference Intakes (DRI) for Ca & Mg

The process of assimilation of Ca and Mg from the water which we consume daily has been demonstrated by work in France (Galan, 2002). The results are reliable and

Table 15.6 Content of Ca & Mg in tap water in large cities of Poland.

	Ca/Mg
Northern Poland:	
Gdańsk:	58–121/9–14
Sopot:	56–100/6–26
Gdynia:	60–92 /7–11
Central Poland:	
Bydgoszcz:	71–117/11–16
Poznań:	81–96 /13–27
Warszawa:	92–104/no data
Zielona Góra:	91–110/9–18
Łódź:	64–72 /5–8
Southern Poland:	
Częstochowa:	68–81 /2–4
Kielce:	62–90 /8–12
Wrocław:	86–93 /11–12
Katowice:	31–69 /4–19
Kraków:	41–104/5–10
Przemyśl (average):	52 /10

Table 15.7 Standards in Europe and Poland: References intakes for Ca – RDA (mg/daily).

Europe*			Poland**		
Age	Females	Males	Age	Females	Males
0–6 months		400	0–6 months	200 (AI)	
6–12 months		600	6–12 months	260 (AI)	
1–5	800	800	1–3	700	700
6–10	1200	1200	4–9	1000	1000
11–24	1500	1500	10–18	1300	1300
25–65	1000	1000	19–50	1000	–
			51–75	1200	–
			19–65	–	1000
>65	1500	1500	66–75	–	1200
			>75	1200	1200
Pregnacy	1500–2000	–	Pregnancy	1300–1000	–
Lactaction	1500–2000	–	Lactation	1300–1000	–

* www.krispin.com; www.lenntech.com
** Institute of Food and Nutrition, 2012, Poland.

Table 15.8 Standards in Europe and Poland: References intakes for Mg – RDA (mg/daily).

Austria, Germany, Switzerland*			Poland**		
Age	Females	Males	Age	Females	Males
1–4		80	0–1	30–70 (AI)	
4–7		120	1–3	80	
7–10		170	4–9	150	
10–13	230	250	10–12	240	240
13–15	310	310	13–18	360	410
15–19	350	400	19–30	310	400
19–25	310	400	31–75	320	420
25–51	300	350			
51–65			>75	320	420
>65					
Pregnancy	310	–	Pregnancy	400–360	–
Lactation	390	–	Lactation	360–320	–

* D-A-CHReferenzwerte fur die Nahstoffzufur; (Grzebisz).
** Institute of Food and Nutrition, 2012, Poland.

unexpected. The French studies have shown that water is not a primary source of Ca and Mg, because these two elements are absorbed only by a small amount from water.

The objective was the assessment of the contribution of mineral water containing different amounts of Ca and Mg to the total dietary intakes of these minerals. The subject of the study was a population of the 424 women, 240 men in 4 groups:

I   regular drinking of Ca-rich (486 mg/l) and Mg-rich mineral water (84 mg/l)
II regular drinking of mineral water: Ca at 202 mg/l and Mg at 36 mg/l

III   drinking of two low-mineralized waters Ca at 9.9–67.6 mg/l and Mg at 1.6 to 2.0 mg/l

IV   drinking of ordinary potable waters.

Depending on the Ca concentration, mineral water may contribute to 25% of the total daily Ca intake. The people who regularly drink highly-mineralised water have a calcium intake significantly higher than the people drinking either low-mineralised or potable water. For Mg the total daily intake varies between 6 and 17% depending on the concentration in the water. People drinking Mg-rich mineral water with medium-mineralisation have a magnesium intake significantly higher then people drinking low-mineralised or potable water. Mineral-rich water may provide an important supplementary contribution to the total Ca and Mg intake.

15.3   CONCLUSIONS

Ca and Mg are (besides sodium) the main macronutrients in groundwater (with the co-existance of Na, Cl, SO_4, HCO_3, iodide, CO_2 and temperature used for:

*   permanent public water supply,
*   provision of bottled waters–for drinking and making beverages,
*   use of certain types of groundwaters for therapeutic purposes–bathing, inhalation, drinking therapy.

Significant Ca and Mg rich waters occur only in highly-mineralised, natural mineral waters such as the HCO_3–Ca–Mg type with free CO_2, or HCO_3–Cl–Ca–Mg/Na type.

Brines with a very high content of Ca and Mg (in the form of chlorides or sulphates) may be considered as therapeutic waters, useful mainly for bathing and inhalation, but unfortunately not for supplementation of calcium and magnesium. In the human diet the main sources of Ca and Mg are dairy products not water.

REFERENCES

Announcement of the Chief Sanitary Inspector of 8th August 2012 (2012) Official Journal MH No. 62, on the publication of a list of waters recognized as NMW.

Database of physical and chemical analyses of therapeutic waters, NMW, spring water, table water and drinking water (2006–2013). Department of Health Resort Materials, NIPH – NIH.

Galan P. et al., (2002) Contribution of mineral waters to dietary calcium and magnesium intake in a French adult population. Journal of the American Dietetic Association. Volume 102, Issue 11, pages 1658–1662.

Geological and Mining Law. Journal of Laws No. 163 pos. 981. 2011.

Grzebisz Witold (2011) Magnesium – food and human health. Journal of Elementology, 16(2), 299–323.

Nutrition standards for the Polish population. Amendment. (2012) Institute of Food & Nutrition, Warsaw.

Regulation of the Minister of Health of 13th April 2006 (2006) Journal of Laws No. 80 pos. 565, on the scope of the research necessary to determine the medicinal properties of natural medicinal (therapeutic) materials and medicinal properties of the climate, their evaluation criteria and the model certificate confirming these properties.

Regulation of the Minister of Health of 20th April 2010 (2010) Journal of Laws No. 72 pos. 466, amending the Regulation on the quality of water intended for human consumption.

Regulation of the Minister of Health of 31st March 2011 (2011) Journal of Laws No. 85 pos.466, of natural mineral water, spring water and table water.

Zachwieja Z., Bartoń H., Fołta M. (2006) Magnesium and calcium in mineral waters and tap water in the light of the Dietary Guidelines. Conference Water for Health.

Chapter 16

A survey of Ca and Mg in thermal springs of West Africa: Implications for drinking and bathing therapy

K'tso Nghargbu[1], Krzysztof Schoeneich[2], Irena Ponikowska[3], Teresa Latour[4] & S. Ayodele Alagbe[2]

[1]*Department of Geology and Mining, Nasarawa State University Keffi, Nigeria*
[2]*Department of Geology, Ahmadu Bello University, Zaria, Nigeria*
[3]*Collegium Medicum Bydgoszcz, Nicholaus Copernicus University in Torun, Poland*
[4]*Department of Health Resort Materials, National Institute of Public Health and Hygiene, Warsaw, Poland*

ABSTRACT

West Africa is located in the tropics with 70% of its territory bounded by the Atlantic shore line. Its potential mineral waters, tourism and the bottled mineral water industry is far from harnessed, hence the current survey of the essential elements such as Ca and Mg in the thirteen most promising medicinal springs of the African sub-region. A comparison of their concentrations in these waters indicate ranges for Ca and Mg suitable for both internal and external cures when compared with concentrations in selected springs used by some Polish health resorts. The concentrations for Ca in West Africa were lowest at 4.81 mg/dm^3 and highest at 110.22 mg/dm^3 while values for Mg had 1.94 mg/dm^3 as lowest and 43.75 mg/dm^3 as highest. The lowest level for the sampled springs in Poland was 144.29 mg/dm^3 and 1503.00 mg/dm^3 for Ca, as against 51.04 mg/dm^3 and 947.90 mg/dm^3 for Mg. While some of the spring waters are meteoric in nature-low mineralised, and hence suitable for bottling as dietary waters, others are moderately mineralised to chlorosodic mineral waters suitable for application in hydrotherapy. 'Possotome' water is the only mineral water brand in Benin Republic, but there are none in Nigeria. This result might perhaps boost the knowledge of medical hydrogeologists, the World Health Organisation, and prospective investors in the bottled mineral water industry in West Africa.

16.1 FIELDWORK

The sampling of selected medicinal springs lasted for about one whole year and included ten countries: Benin, Cote D'Ivoire, Gambia, Ghana, Liberia, Niger, Nigeria, Senegal, Sierra Leone, and Togo.

Springs of interest were located in three countries namely: Benin, Niger and Nigeria (Figure 16.1 and Table 16.1).

A Pictorial View of the springs is displayed on plates 1 to 8.

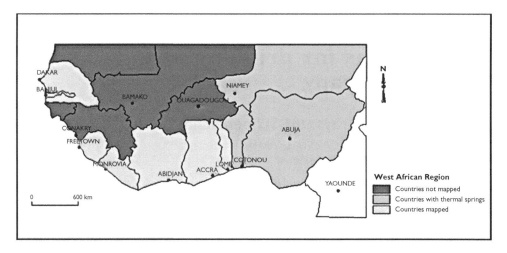

Figure 16.1 Countries mapped with indication of those with medicinal springs.

16.2 METHODOLOGY

The sample water was collected into 2 l sample bottles and then sent to the health resort materials laboratory of the Polish Institute of Health and Hygiene in Warsaw for analysis using a combination of techniques. The Electrometric method was used to determine electrical conductivity, pH value, iodide and fluoride content. The spectrophotometric was used for the determination of: ammonium, Fe, Mg, NO_3, nitrite, silicate, bromate, and sulphate content. The flame photometric method was used for the determination of Na, and K. Finally, the complexometric titration was used to determinate Ca, and Mg, the argentometric titration to determinate Cl, while the acidimetric titration was used in the determination of bicarbonate.

16.3 GEOLOGY OF WEST AFRICA

The principal geological units of West Africa west of the Guinea Rise, were formed along the southern half of the West African Craton, sometimes also called the Man Shield (Figure 16.2). The narrow Pan African Belt of the Rokelids on its western boundary is associated with some marginal reactivation of the Archaean rock. North of the Craton lie Infra-Cambrian to Lower Paleozoic sediments of the great Taoudeni Basin, obscured in the east by continental Tertiary to Quaternary deposits. The Bove Basin in the south west is occupied by Lower Palaeozoic sediments which interrupt the continuity of the Rokelide Belt with the Mauritanide belt, which forms the eastern boundary of the much younger (Mesozoic-Tertiary) Senegal Basin.

Table 16.1 Spring locations and some physical parameters of water.

Spring name	Country	Coordinates						Some physical parameters of water			
		Latitude (N)			Longitude (E)						
		Degrees	Minutes	Seconds	Degrees	Minutes	Seconds	pH	EC (mS/cm)	TDS (mg/l)	Temperature (°C)
Tafadek	Niger	17	23	19.2	7	57	27.5	7.74	1.09	620	46
Possotome	Benin	6	31	21.7	1	58	12.6	7.74	0.82	520	43
Zagnanado	Benin	7	13	10.3	2	23	55.2	6.08	0.02	10	29.6
Atchabita	Benin	6	53	39.3	2	27	14.6	7.23	0.56	330	40.2
Hetim-Sota	Benin	6	35	15	2	30	17	7.23	1.07	600	47
Ruwan Zafi (Adm)	Nigeria	9	28	46	11	30	6.9	7.23	0.51	310	44.2
Ikogosi	Nigeria	7	35	40.8	4	58	50.3	7.23	0.08	40	35.6
Wikki	Nigeria	9	45	11.1	10	30	40.3	7.24	0.01	10	32.9
Ruwan Zafi (Kd)	Nigeria	10	25	35.9	8	30	49.8	7.22	10.89	6270	46.5
Akiri	Nigeria	8	22	51.3	9	20	10	7.23	11.9	6770	35.8
Ruwan Gishiri	Nigeria	8	26	55.5	9	4	29.6	7.23	16.02	9110	39.5
Ruwan Zafi, Awe	Nigeria	8	6	1.06	9	8	2.35	7.23	14.16	8060	34
Tangarahu	Nigeria	8	7	44.4	9	29	58.6	7.23	16.24	9230	42.7
Bitrus	Nigeria	8	11	25.3	9	2	44.9	7.23			

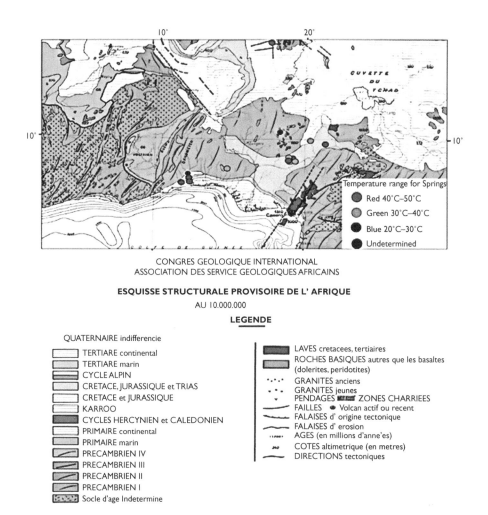

Figure 16.2 Springs discussed in this research superimposed on geological map of West Africa.

16.3.1   Geology of the countries with hot springs

These countries are Benin Republic, Niger and Nigeria. They all have large areas of basement (metamorphic), igneous, and sedimentary rocks. In Nigeria for instance, the basement rocks occupy about 40% of the entire country, the igneous rocks about 10% while the remainder is covered by sedimentary rocks. The north central parts and the south western parts of the country are covered by basement rocks while the north western and north eastern parts are dominated by sedimentary basins, which also run through parts of the north east, north central, and the entire south east and southern parts of the country with the youngest Tertiary to Recent sedimentary beds found within the Niger Delta area.

In the case of Benin, 70% of the area is covered by basement rocks while the sedimentary rocks account for the remaining 30%. The northern region is dominated by crystalline rocks of the Neoproterozoic-Dahomeyide Orogeny, whereas the southern region consists of sedimentary rocks ranging from Recent to Cretaceous age. In the north eastern part of the country alluvial rocks of Neogene predominate.

For Niger Republic, most of the country is covered by sedimentary rocks from the Cretaceous to Quaternary age. About 10% of the Sahara desert area is made up of the youngest sets of Younger Granite rocks in the Agadez region. Rocks of the Pan-African age also exist.

Table 16.2 Composition of selected bottled mineral waters in Poland.

Name of water	Anions [mg/l]						Cations [mg/l]				
	HCO_3^-	Cl^-	SO_4^{2-}	F^-	J^-	NO_3^-	Na^+	K^+	Ca^{2+}	Mg^{2+}	$Fe^{2+/3+}$
Kropla Beskidu non carbonated	186.70	3.19	43.62	–	–	–	11.10	1.00	44.09	17.01	–
Artic carbonated	260.10	–	26.54	0.27	<0.02	<0.20	8.12	1.35	74.15	13.37	1.15
Jurajska non carbonated	329.90	7.80	40.50	0.40	–	–	10.00	2.20	66.10	32.80	–
Polaris non carbonated	432.7	2.50	–	0.23	–	–	11.25	2.34	102.20	16.00	–
Kropla Minerałów low carbonated CO_2	918.20	6.40	17.50	0.20	–	–	80.60	6.00	159.30	50.40	–
Żywiec Zdrój carbonated	201.50	–	–	–	–	–	4.00	–	62.12	6.08	–
Górska Natura carbonated	–	<5.00	–	<0.10	–	–	1.05	–	24.60	6.47	–
Górska Natura non carbonated	104.00	<5.00	<25.00	<0.10	–	6.00	1.04	1.93	25.20	6.47	<0.01
Kuracjusz Beskidzki non carbonated	384.40	7.00	44.85	0.08	–	–	100.00	1.00	41.10	11.54	0.20

Table 16.3 Concentration levels of Ca in the sampled springs of west Africa against those of selected bottled mineral waters produced in Poland.

S/N	West African springs	Ca(mg/l)	Bottled mineral water brands in Poland	Ca (mg/l)
1.	Tafadek	21.043	Kropla Beskidu non carbonated	44.09
2.	Zagnanado	4.81	Arctic carbonated	74.15
3.	Possotome	20.04	Jurajska non carbonated	66.1
4.	Hetim-Sota	7.21	Polaris-Niegazowana	102.2
5.	Atchabita-Bonou	15.63	Kropla Minerałów low carbonated Co_2	159.3
6.	Ikogosi	8.82	Żywiec Zdrój carbonated	62.12
7.	Ruwan Zafi, Numan	9.62	Górska Natura carbonated	24.60
8.	Wikki	6.01	Górska Natura non carbonated	25.20
9.	Akiri	110.22	Kuracjusz Beskidski non carbonated	41.10
10.	Ruwan Gishiri	108.22		
11.	Ruwan Zafi, Awe	40.08		
12.	Tangarahu	40.10		
13.	Bitrus	44.10		

Table 16.4 Concentration levels of magnesium in the sampled springs of West Africa against those of selected bottled mineral waters produced in Poland.

S/N	West African springs	Magnesium in content (mg/l)	Bottled mineral water brands in Poland	Magnesium content (mg/l)
1.	Tafadek	4.25	Kropla Beskidu non carbonated	17.01
2.	Zagnanado	1.94	Arctic carbonated	13.37
3.	Possotome	18.47	Jurajska non carbonated	32.8
4.	Hetim-Sota	2.92	Polaris non carbonated	16.00
5.	Atchabita-Bonou	15.80	Kropla Minerałów low carbonated CO_2	50.4
6.	Ikogosi	5.35	Żywiec Zdrój carbonated	6.08
7.	Ruwan Zafi, Numan	2.43	Górska Natura carbonated	6.47
8.	Wikki	1.94	Górska Natura non carbonated	6.47
9.	Akiri	20.66	Kuracjusz Beskidski non carbonated	11.54
10.	Ruwan Gishiri	20.05		
11.	Ruwan Zafi, Awe	43.75		
12.	Tangarahu	36.46		
13.	Bitrus	38.90		

Table 16.5 Information on mineral/medicinal water boreholes/springs in Poland analysed between June and July 2008 (Source: Health Resorts, Solec, Swinoujscie, and Ciechocinek).

S/N	Name of borehole/ spring	Locality	District	Province	Calcium concentration in mg/l	Magnesium concentration in mg/l
1.	Malina	Wetnin	Solec Zdroj	Holy Cross	1102.20	947.90
2.	XXX – Lecia (Nr IVa)	Swinoujscie	Swinoujscie	Zachodniopomorskie	961.92	437.49
3.	Teresa (Nr VI)	Swinoujscie	Swinoujscie	Zachodniopomorskie	881.76	425.34
4.	Jantar (Nr V)	Swinoujscie	Swinoujscie	Zachodniopomorskie	821.64	449.64
5.	Nr 14	Ciechocinek	Ciechocinek	Kujawskopomorskie	1322.64	486.10
6.	Nr 16	Ciechocinek	Ciechocinek	Kujawskopomorskie	1503.00	486.10
7.	Nr 19a	Ciechocinek	Ciechocinek	Kujawskopomorskie	144.29	51.04

Table 16.6 Highest and lowest concentration levels of calcium and magnesium in the sampled springs of West Africa analysed at Health Resort Materials Laboratory, Poznan, Poland in September 2011.

S/N	Name of spring	Country	Calcium concentration in mg/l	Magnesium concentration in mg/l
1.	Akiri	Nigeria	110.22	–
2.	Ruwan Zafi, Awe	Nigeria	–	43.75
3.	Zagnanado	Benin Republic	4.81	1.94
4.	Wikki	Nigeria	–	1.94

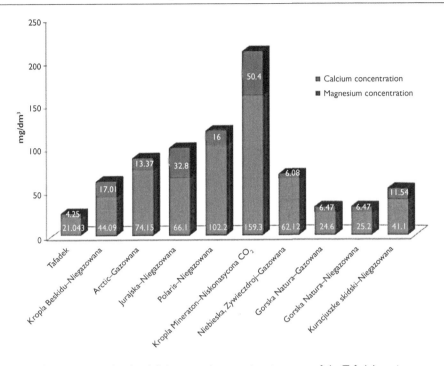

Figure 16.3 Concentration levels of Calcium and magnesium in water of the Tafadek spring compared with those of some bottled waters produced in Poland.

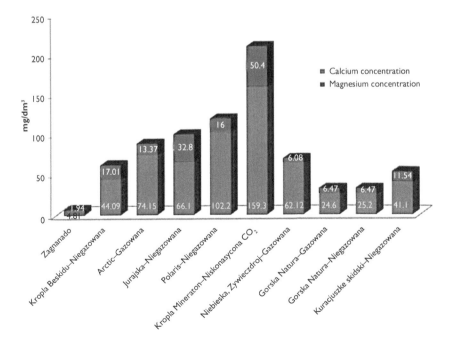

Figure 16.4 Concentration levels of Calcium and magnesium in water of the Zagnanado spring compared with those of some bottled waters produced in Poland.

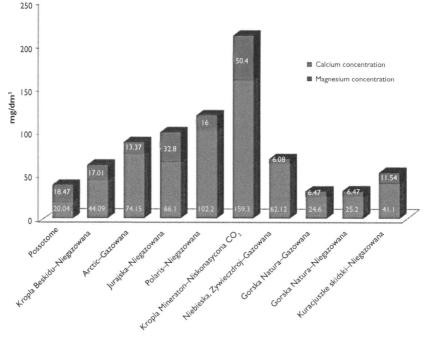

Figure 16.5 Concentration levels of Calcium and magnesium in water of the Possotome spring compared with those of some bottled waters produced in Poland.

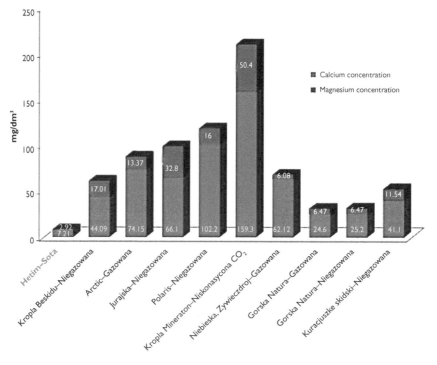

Figure 16.6 Concentration levels of Calcium and magnesium in water of the Hetim–Sota spring compared with those of some bottled waters produced in Poland.

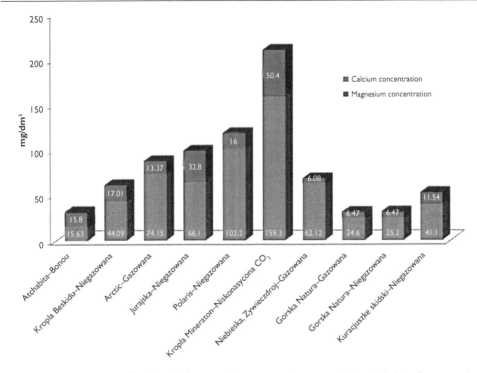

Figure 16.7 Concentration levels of Calcium and magnesium in water of the Atchabita–Bonou spring compared with those of some bottled waters produced in Poland.

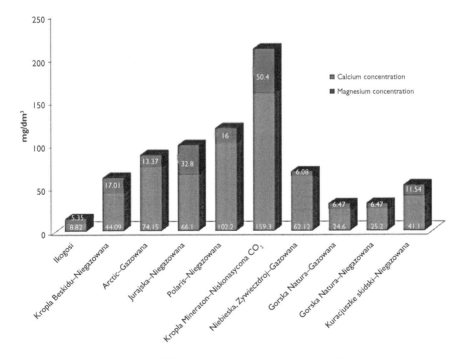

Figure 16.8 Concentration levels of Calcium and magnesium in water of the Ikogosi spring compared with those of some bottled waters produced in Poland.

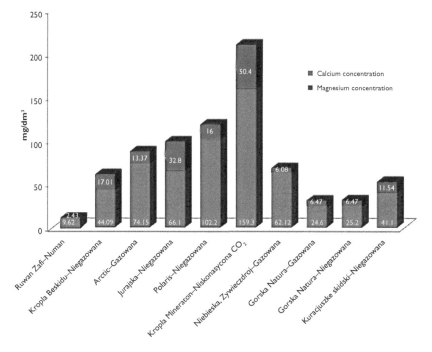

Figure 16.9 Concentration levels of Calcium and magnesium in water of the Ruwan Zafi–Numan spring compared with those of some bottled waters produced in Poland.

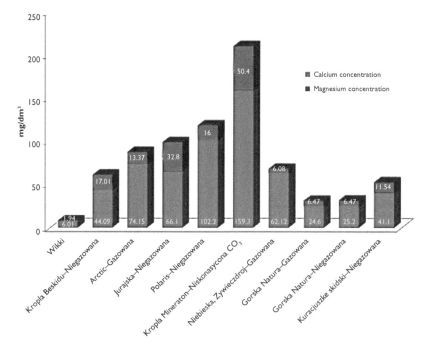

Figure 16.10 Concentration levels of Calcium and magnesium in water of the Wikki spring compared with those of some bottled waters produced in Poland.

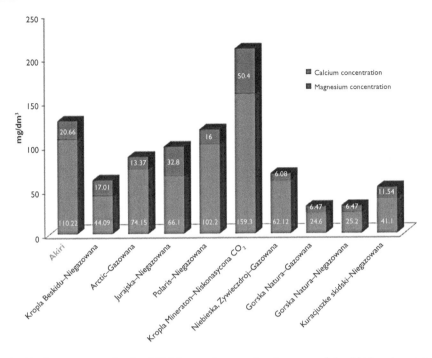

Figure 16.11 Concentration levels of Calcium and magnesium in water of the Akiri spring compared with those of some bottled waters produced in Poland.

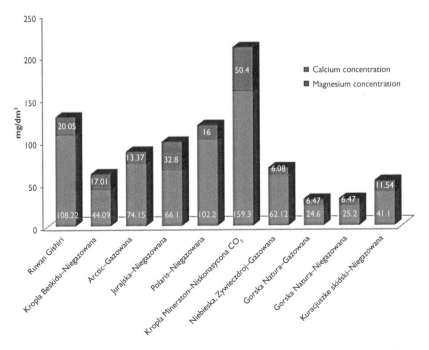

Figure 16.12 Concentration levels of Calcium and magnesium in water of the Ruwan Gishiri spring compared with those of some bottled waters produced in Poland.

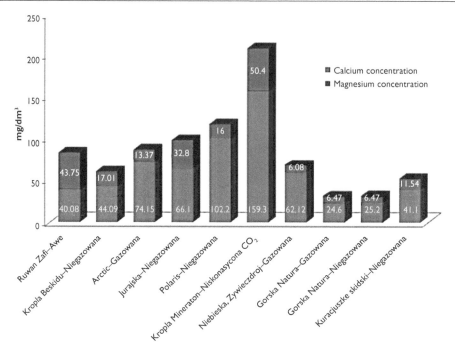

Figure 16.13 Concentration levels of Calcium and magnesium in water of the Ruwan Zafi–Awe spring compared with those of some bottled waters produced in Poland.

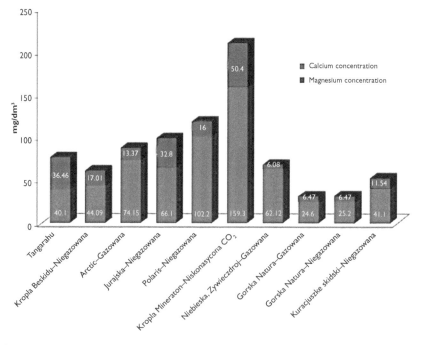

Figure 16.14 Concentration levels of Calcium and magnesium in water of the Tangarahu spring compared with those of some bottled waters produced in Poland.

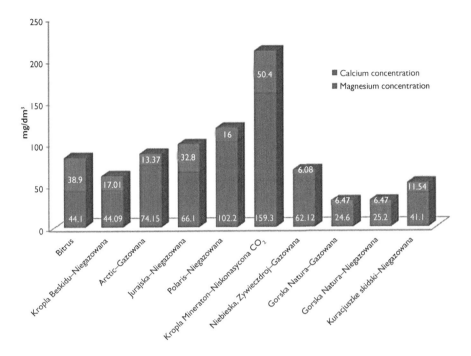

Figure 16.15 Concentration levels of Calcium and magnesium in water of the Bitrus spring compared with those of some bottled waters produced in Poland.

16.4 RESULTS AND DISCUSSION

In order to draw a reliable conclusion on the desirability or otherwise of the concentrations of Ca and Mg in the selected West African springs for both internal and external use, the results obtained from their analyses were compared with those of selected bottled mineral concentration levels whose production is done in Poland as shown in Table 16.2.

Tables 16.3 to 16.6 provide a tabular comparison of Ca and Mg concentrations in the sampled springs and those in selected bottled mineral waters produced in Poland and details are shown in Figures 16.3 to 16.15.

16.5 CONCLUSION

Based on the tabular and statistical comparison of the result of analyses of concentration levels in the West African Springs with those in Poland including levels in bottled mineral waters, it is apparent that Ca and Mg in the West African Springs are at a safe level for internal cures as dietary waters as well as essential macro-nutrients in waters needed for external cures as indicated by a favourable comparison of Ca in Akiri and Mg in Ruwan Zafi, Awe (Table 16.5) with concentration levels of borehole Number 19a Ciechocinek (See Table 16.4). For now, only the Possotome Spring in

Benin Republic which flows through sandstone, limestone and siltstone is exploited as a source of dietary water in West Africa. The spring waters do not attract tourists. This is as a result of setbacks in the tourism sector in West Africa due to insecurity, political instability, poor economic policies, as well as massive degradation of the environment. However, two of the springs are mapped as tourist sites, Ikogosi Spring and the Wikki Spring, and both are in Nigeria.

ACKNOWLEDGEMENTS

Staff and management of Health Resorts, Solec, Ciechocinek, and Swinoujscie are thanked for supplying the needed analytical results from their boreholes. The Polish Geological Institute, Upper Silesian Branch is also appreciated for providing the platform for this work.

REFERENCES

Finkelman R.B. (2006) Health Benefits of Geologic Materials and Geologic Processes. *International Journal of Environmental Research and Public Health* 3(4), 338–342.
Food and Agricultural Organisation Year Book, (1981).
Jahnke H.E. (1982) Livestock Production Systems in Livestock Development in Tropical Africa. *Nature* 182, 1097–98.
Jarrett H.R. (1980) The Remaining Sahel Territories: Mauritania, Mali, Upper Volta and Niger. In: *A Geography of West Africa* (Evans Brothers, eds). BAS Printers Limited, Over Wallop, Hampshire, 140–142.
Ponikowska I. (2009) Clinical Basis Balneology and Physical Medicine. In: *Health and Rest* (Zbigniew Franczukowski *et al.* eds.). Wydawnictwo Mirex, Bydgoszcz, Poland. 41–54.
Ponikowska I. & Ferson D. (2009) Nowoczesna Medycyna Uzdrowiskowa. Medi Press, Warsaw, Poland. 1–365.

Internet Resources accessed between May and August 2012:
Balneotherapy and Balneology; *The Science and Art of Mineral Water Therapy (2006)*: www.eytonsearth.org/balneology-balneotherapy.php.
Climatic Indexes of West Africa: http://www.climatemps.com/graph/.
Cultures of West Africa: http://www.everyculture.com.
Healing Waters (2006): www.healingwaters.htm.
Health Resort, Busko-Zdroj: www.umig.busko.pl.
Hotel Medical Spa: http//:www.malinowyzdroj.pl; Hotel Medical Spa Health Information Booklet.
Hydrogeology in Relation to Other Fields (2004): www.answers.com.
Medical Geology, (2011): An IMGA Poster Presented at the Fourth International Medical Geology Conference, Bari, Italy. http://www.medicalgeology.org.
The Hotsprings of Iceland: www.Youtube.com/iceland_hotsprings).
The Medical Geology Revolution-The Evolution of an IUGS Initiative (2005): www iugs.org.
Therapeutic Radioenergic Hot Mineral Springs (2006): www.hotmineralsprings.htm.

Author index

Subject index

Series IAH-selected papers

14. Advances in Subsurface Pollution of Porous Media: Indicators, Processes
    and Modelling
    Edited by: Lucila Candela, Iñaki Vadillo and Francisco Javier Elorza
    2008, ISBN Hb: 978-0-415-47690-4

15. Groundwater Governance in the Indo-Gangetic and Yellow River Basins –
    Realities and Challenges
    Edited by: Aditi Mukherji, Karen G. Villholth, Bharat R. Sharma
                and Jinxia Wang
    2009, ISBN Hb: 978-0-415-46580-9

16. Groundwater Response to Changing Climate
    Edited by: Makoto Taniguchi and Ian P. Holman
    2010, ISBN Hb: 978-0-415-54493-1

17. Groundwater Quality Sustainability
    Edited by: Piotr Maloszewski, Stanisław Witczak and Grzegorz Malina
    2013, ISBN Hb: 978-0-415-69841-2

18. Groundwater and Ecosystems
    Edited by: Luís Ribeiro, Tibor Y. Stigter, António Chambel, M. Teresa Condesso
                de Melo, José Paulo Monteiro and Albino Medeiros
    2013, ISBN Hb: 978-1-138-00033-9

19. Assessing and Managing Groundwater in Different Environments
    Edited by: Jude Cobbing, Shafick Adams, Ingrid Dennis and Kornelius Riemann
    2013, ISBN Hb: 978-1-138-00100-8

20. Fractured Rock Hydrogeology
    Edited by: John M. Sharp
    2014, ISBN Hb: 978-1-138-00159-6